Wildlife Behavior and Conservation

Richard H. Yahner

Wildlife Behavior
and Conservation

Richard H. Yahner
The Pennsylvania State University
University Park, PA 16802, USA

ISBN 978-1-4614-1515-2   e-ISBN 978-1-4614-1518-3
DOI 10.1007/978-1-4614-1518-3
Springer New York Dordrecht Heidelberg London

Library of Congress Control Number: 2011943350

© Springer Science+Business Media, LLC 2012
All rights reserved. This work may not be translated or copied in whole or in part without the written permission of the publisher (Springer Science+Business Media, LLC, 233 Spring Street, New York, NY 10013, USA), except for brief excerpts in connection with reviews or scholarly analysis. Use in connection with any form of information storage and retrieval, electronic adaptation, computer software, or by similar or dissimilar methodology now known or hereafter developed is forbidden.
The use in this publication of trade names, trademarks, service marks, and similar terms, even if they are not identified as such, is not to be taken as an expression of opinion as to whether or not they are subject to proprietary rights.

Printed on acid-free paper

Springer is part of Springer Science+Business Media (www.springer.com)

I dedicate this book to my three closest friends: Rich, Tom, and Audrey.

Acknowledgment

This book is a product of my long-standing interest in wildlife conservation and behavior. Research on behavior over the years has been funded by the Pennsylvania Agricultural Research Center, the Pennsylvania Game Commission, the Pennsylvania Wild Resource Conservation Fund, the US Forest Service, the National Park Service, the US Environmental Protection Agency, the Hammermill and International Paper Companies, the Hawk Mountain Sanctuary, the National Rifle Association, the Western Pennsylvania Conservancy, and several utilities, including GPU, Penelec (First Energy), PECO, and ComED. I also am grateful to personnel within these agencies and to my numerous graduate students at both The University of Minnesota and The Pennsylvania State University. I thank J. Grant for the illustrations.

About the Author

Richard H. Yahner is Professor of Wildlife Conservation in the School of Forest Resources at The Pennsylvania State University. Previously, he was Assistant Director for Outreach in the School of Forest Resources, Associate Dean of the Graduate School at Penn State, Assistant Professor of Wildlife at the University of Minnesota, and Postdoctoral Fellow at the Smithsonian Institution. He has taught numerous courses in wildlife, including Ecology and Behavior. He has published over 300 publications on wildlife conservation, ecology, and behavior.

Contents

1 Comparative Psychology Versus Ethology 1
   1.1   Behavior and Wildlife Management 1
   1.2   Comparative Psychology Versus Ethology 2

2 Genetics and Other Mechanisms Affecting Behavior 5
   2.1   Introduction 5
   2.2   Diversity in Behavior Diversity 5
   2.3   Sociobiology and Behavioral Ecology 6
   2.4   Social Organization or Social System 6
   2.5   Social Units 6
   2.6   Matriarchal Social Units 9
   2.7   Plasticity in Social Unit Size 9
   2.8   What Do Animals Learn in a Social Unit? 10
   2.9   Cultural Transmission of Learning 10
   2.10 Ultimate Versus Proximate Factors in Wildlife Behavior 11
   2.11 Hormones and Proximate Factors 12
   2.12 The Nervous System, Biochemistry, and Behavior 13

3 Mate-acquisition and Parental-Care Systems 15
   3.1   Courtship and Mating Systems 15
   3.2   Monogamy 15
   3.3   Polygamy 17
   3.4   Promiscuity and Other Mating Systems 20
   3.5   Mating and Mechanisms of Mating Interference 22
   3.6   Mate Choice 22

4 Mating Systems and Parental Care 25
   4.1   Courtship 25
   4.2   Parental-Care Systems 25
   4.3   Altruism and Parental Care 27
   4.4   Brood Parasitism and Parental Care 28
   4.5   Brown-Headed Cowbird 29

	4.6	Altricial and Precocial Young	31
	4.7	Lactation in Mammals	32
	4.8	Pouches, Parental Care, Locomotion, and Altricial Versus Precocial Young	32
	4.9	Duration of Parental Care and the Timing of Dispersal	33
5	**Dispersal and Corridors**		**35**
	5.1	Timing of Dispersal	35
	5.2	Reasons to Disperse or Not to Disperse	35
		5.2.1 Corridors in the Landscape	37
		5.2.2 Landscape Linkages or Megacorridors	39
6	**Food-Acquisition Systems**		**41**
	6.1	General Comments	41
	6.2	Wildlife and the Prey Rat Race	42
	6.3	Optimal Foraging Theory	43
	6.4	Central Place Foraging and Hoarding of Food	46
	6.5	Constraints on Optimal Foraging: Predation and Competition	48
	6.6	Foraging and Group Life	49
	6.7	Predation and Prey Distribution	50
	6.8	Humans as Prey	51
	6.9	Conservation and Warfare	53
7	**Additional Adaptations Against Predation**		**55**
	7.1	Some Less-Direct Adaptations	55
	7.2	Warning Coloration	58
	7.3	Mimicry	59
	7.4	Playing Possum and Enhancement	61
	7.5	Weaponry in Animals	62
8	**Habitat Selection**		**65**
	8.1	Selection Versus Use of a Substrate or Habitat	65
	8.2	Testing Habitat Selection Versus Use	66
	8.3	Habitat Selection and Urban Wildlife	66
	8.4	Habitat Selection and Wildlife Recolonization	67
	8.5	Some Factors Affecting Habitat Selection	67
	8.6	Other Factors Affecting Habitat Selection	69
		8.6.1 Ambient Temperature	69
		8.6.2 Acid Deposition	70
		8.6.3 Population Density	70
		8.6.4 Tradition	71
	8.7	Is Habitat Selection Learned?	72
	8.8	Role of Resources: Generalist Versus Specialist	73
	8.9	Some Problems in Quantifying Habitat Selection	75

9 Home Range and Homing ... 77
- 9.1 Some Comments About Home Range ... 77
- 9.2 Relationship Between Home Range and Dispersion Patterns ... 77
- 9.3 Homing in Wildlife ... 78
- 9.4 Quantifying a Home Range ... 79
- 9.5 How Large Is a Home Range ... 79
- 9.6 What Is the Shape of a Home Range? ... 80

10 Spacing Mechanisms ... 83
- 10.1 Introduction to Territoriality ... 83
- 10.2 How is a Territory Defended? ... 84
- 10.3 Problems in Defining a Territory and Ontogeny of Territoriality ... 86
- 10.4 Is Territoriality Genetically Programmed and Static? ... 87
- 10.5 How Do Animals Know Neighbors? ... 87
- 10.6 How Large Is a Territory? ... 88
- 10.7 The Evolution of Territoriality ... 88
- 10.8 Interspecific Territoriality ... 90
- 10.9 Introduction to Defense of Individual Distance ... 91
- 10.10 Other Differences in Defense of Individual Distance ... 92

11 Dominance Hierarchies ... 95
- 11.1 Introduction to Dominance Hierarchies ... 95
- 11.2 Advantages of Hierarchies ... 97
- 11.3 Why Be Subordinate and Stay in a Group? ... 97
- 11.4 How Is Dominance Measured? ... 98
- 11.5 Maintenance and Establishment of Hierarchies ... 98
- 11.6 Is a Dominance Hierarchy Better than Other Mechanisms? ... 100

12 Communication ... 101
- 12.1 Introduction to Communication ... 101
- 12.2 Introduction to Visual Communication ... 102
- 12.3 Roles of Visual Communication ... 103
- 12.4 Spatial and Temporal Contexts of Visual Threat Displays ... 104
- 12.5 Types of Threat Displays and Decision Making by Animals ... 105
- 12.6 Evolutionary Origin of Displays ... 106
- 12.7 Evolution of Tusks in Walrus ... 106
- 12.8 Evolution of Antlers and Horns ... 107
- 12.9 Artificial Night Lighting ... 108
- 12.10 Possible Effects of Ecological Light Pollution on Wildlife ... 109

13 Olfactory Communication ... 113
- 13.1 Introduction to Olfactory Communication ... 113
- 13.2 Olfaction in Various Wildlife Species ... 113
- 13.3 Olfaction as Alarm Signals ... 115
- 13.4 Buck Rubs ... 115
- 13.5 Buck Scrapes ... 116
- 13.6 A Universal Trend ... 116
- 13.7 Cheek Rubbing ... 116
- 13.8 Olfaction and Infanticide ... 117
- 13.9 Scent-Marking in Canids ... 117
- 13.10 Mud Piles in Beaver ... 118
- 13.11 Use of Scent Stations in Mammalian Research ... 119

14 Auditory Communication ... 121
- 14.1 Introduction to Auditory Communication ... 121
- 14.2 Barking in Forest Deer ... 121
- 14.3 Barking and Domestication in Domestic Dogs ... 122
- 14.4 Vocalizations in Wildlife ... 123
- 14.5 Bird Vocalizations ... 124
- 14.6 How Do Birds Identify Songs? ... 126
- 14.7 Song Repertoires in Birds ... 126
- 14.8 Mimicry in Birds ... 127
- 14.9 Sound-Producing Mechanisms ... 127
- 14.10 Effects of Sound on Terrestrial Animals ... 129

15 Ultrasounds and Other Types of Communication ... 131
- 15.1 Introduction ... 131
- 15.2 Echolocation in Bats and Birds ... 131
- 15.3 Echolocation in Shrews ... 132
- 15.4 Sounds in Pinnipeds, Toothed Whales, and African Elephants ... 132
- 15.5 Introduction to Tactile and Electrical Communication ... 134
- 15.6 Tactile Communication ... 134
- 15.7 Play Behavior ... 135
- 15.8 Seismic Communication ... 135
- 15.9 Electrical Communication ... 136

16 Winter Strategies ... 139
- 16.1 Introduction to Winter Strategies ... 139
- 16.2 Hibernation ... 140
- 16.3 Hibernation in the Woodchuck ... 140
- 16.4 Hibernation in Bats ... 141
- 16.5 Winter Lethargy in Black Bear ... 141
- 16.6 Torpor in Eastern Chipmunks ... 142
- 16.7 Daily Torpor in Aerial Animals ... 142

17 Migration, Orientation, and Navigation ... 145
- 17.1 Introduction ... 145
- 17.2 Advantage of Migration Over Dormancy ... 145
- 17.3 Distances of Migration ... 146
- 17.4 Energy Stores for Migration ... 147
- 17.5 Timing of Migration ... 147
- 17.6 Navigational Routes ... 149
- 17.7 Cues Used During Navigation ... 149
- 17.8 Learning to Navigate ... 150

18 Competition ... 151
- 18.1 Introduction ... 151
- 18.2 Interspecific Competition ... 152

References ... 155

Index ... 165

List of Figures

Fig. 1.1 The brown bear, which is also called the grizzly bear, is a large bear that ranges across most of North America (west of the Mississippi River) and in northern Europe and Asia. Infanticide, or the killing of young, has been widely studied in humans, but it also occurs in brown bear and other animals ranging from rotifers to other mammals 2

Fig. 1.2 An American psychologist, B.F. Skinner, invented an operant conditioning chamber, which measured the response of organisms (typically, rats) to a lever that provided food to the organism 3

Fig. 2.1 The woodchuck, or groundhog, is the only marmot in eastern North America. Unlike its western counterparts, the woodchuck is solitary; western marmots tend to be colonial... 7

Fig. 2.2 The sandwich tern is a seabird, in which males plunge for fish and offer the fish caught to females as part of the courtship. Young sandwich terns learn food items from both parents via observational learning 12

Fig. 3.1 The yellow warbler is very wide-spread, extending its range throughout North America and as far south as central Mexico... 16

Fig. 3.2 Fur seals consist of several species, with the northern fur seals occurring in the North Pacific and most other species occurring south of the equator. Typically, fur seals form assemblages each year during summer months to give birth and breed. All species are polygynous (a dominant male reproducing with more than one female) 18

Fig. 4.1	The spotted hyena, or laughing hyena, occurs today only of the south of the Sahara, except in the basin of the Congo. Hyenas live in large matriarch groups, or clan, which may comprise 80 individuals. Hyenas are not scavengers, but actively hunt large mammals	26
Fig. 4.2	The shiny cowbird of South America, although it may breed in southern Florida. Like other cowbirds, the shiny cowbird is a brood parasite	29
Fig. 5.1	The gray wolf, or timber wolf, is very social, living in packs, which often are closely related. The gray wolf shares a common ancestry with the domestic dog	36
Fig. 5.2	The gypsy moth is a forest pest, from Europe and Asia, having escaped into the USA about 1860. Egg masses of gypsy moths are typically placed on branches and trunks of trees. Egg masses are buff colored when first laid but may become bleached over winter	37
Fig. 6.1	The great horned owl has females that are larger than males. Its "horns" are not ears or horns, but tufts of feathers. It is believed to occur from Nova Scotia to east Texas and to Minnesota and southward to South America. Great horned owls breed early in late January or early February	44
Fig. 6.2	The wood stork, or sometimes called the wood ibis, is the only stork that breeds in North America. It extends its range throughout South America, in tropical and subtropical regions	46
Fig. 7.1	The monarch is perhaps the best known of the butterflies in North America; it is often called the milkweed butterfly. Besides occurring in North America, monarchs are found in Europe and New Zealand. It is famous for its migration in the Americas, especially to Mexico, which may span 3–4 generations	56
Fig. 7.2	The fox squirrel is closely related to gray squirrels, but fox squirrels are considered the most polymorphic mammal. In the southeastern USA, for example, the color of fox squirrels has different percentages of gray and black on dorsal surfaces	57
Fig. 8.1	The spotted salamander of eastern North America. It is considered as a mole salamander	71
Fig. 8.2	The northern river otter, or North American river otter, is found only in (endemic) North America. It is a weasel, which is well-adapted to aquatic habitats	73

List of Figures

Fig. 9.1	The common raccoon is native to North America. Habitats of the common raccoon include deciduous and mixed forests, but raccoons are now found in urban and mountainous, and coastal areas. The common raccoon was introduced into Europe and Asia (Japan)	78
Fig. 9.2	The eastern cottontail is among about 16 species in North and South America. It closely resembles the wild European rabbit	81
Fig. 10.1	The northern flicker is a woodpecker native to most of North America and is a woodpecker that migrates. It uses diving and cavity blocking to defend its nest and territory	85
Fig. 10.2	The beaver, or North American beaver, is the only beaver of North America. It was introduced to South America. The only other living (extant) beaver is the European beaver of northern Europe and Asia	86
Fig. 10.3	The brown-headed nuthatch occurs in pine forests of the southeastern USA	92
Fig. 11.1	The chicken was domesticated, probably over 10,000 years ago in Asia. It is now a primary source of food and eggs for humans	96
Fig. 11.2	The eastern chipmunk is a small squirrel of eastern North America. It is solitary (except during its mating season, a female with young), diurnal, and lives in an extensive burrow system	96
Fig. 12.1	Homo habilis, which means "handy man," is closely related to modern humans, but is least similar to that species, with short (less than half the size) and disproportionately long arms	102
Fig. 12.2	The white-tailed deer has expanded its range and now occurs in all 48 states (except perhaps Utah) in the USA, Canada, and as far south as Peru. This deer has been introduced into New Zealand and some European countries	103
Fig. 12.3	The name mussel is used for members of several families of clams of bivalves from fresh and saltwater habitats, but most commonly used in reference to edible bivalves of saltwater habitats. One species, Ligumia nasuta, attracts fish using the display of a white spot on papillae that moves on the mantle of the female mussel	105

Fig. 13.1	A fish lure is an object that often is tied to the end of a fishing line and is meant resemble and move like fish prey. Because fish have a good sense of smell, many lures smell like fish prey	114
Fig. 13.2	The raccoon is usually nocturnal (active at night) and it has two distinctive features, a facial mask and very dexterous forefeet. Raccoons typically are solitary	117
Fig. 14.1	Reeves' muntjac, or Chinese muntjac, is given its common name from John Reeves, who was an Assistant Inspector of Tea for the British East India Company in 1812, and because this species occurs in China	122
Fig. 14.2	The ovenbird is a small, ground-nesting bird of North America. This songbird migrates south as a winter strategy	125
Fig. 15.1	The little brown myotis, or little brown bat, is common in North America. As a winter strategy, this species migrates	132
Fig. 15.2	The African elephant is common to most people. Both male and female have tusks, whereas in the other well-known elephant, the Asiatic or Indian elephant, only males have tusks	133
Fig. 15.3	The collared peccary is pig-like, but it is not in the pig family. Peccaries are often called javelinas. Collared peccary represents a major prey of the large cat, the jaguar, where the two are sympatric (have overlapping distributions)	134
Fig. 15.4	The white-lipped tree frog, or giant tree frog, is native to Australia and is the largest known tree frog	135
Fig. 16.1	The desert tortoise is an ectotherm native to deserts of southwestern USA and northern Mexico (Mojave and Sonoran)	140
Fig. 16.2	The ruby-throated hummingbird is solitary and migratory, and it has a breeding range throughout most of eastern North America	142
Fig. 17.1	The wildebeest, or gnu, is a native to Africa. They are grassland ungulates (animals with hooves); they are well-known for their annual migrations to fresh grass, especially across the Serengeti plains of the national park	146
Fig. 17.2	Mountain sheep, or bighorn sheep, have males with big curving horns. Bighorn sheep are native to North America, having crossed into this continent via the Bering Land Bridge from Siberia	147

List of Figures

Fig. 18.1 Besides humans, the macaques are very widespread, ranging from Asia to Africa, with as many as 22 species. The best known macaque is the Rhesus macaque or monkey 152

Fig. 18.2 The Sika deer, the Japanese spotted deer, is native to Asia and has been introduced into the USA and other countries. The Sika deer retains its spots throughout its lifetime.................. 153

Chapter 1
Comparative Psychology Versus Ethology

1.1 Behavior and Wildlife Management

As humans, we consistently judge the behavior of animals (e.g., that pet is cute) or fellow humans (e.g., that new neighbor is friendly, etc.) based on their appearance and how they act from our perspective. In other words, we seldom look at the ecology of a pet or human or, for that matter, of an animal in the wild. In contrast, wildlife managers typically look at the ecology of an animal or view the animal as being part of a population; management of a species is based on populations rather than on individuals; wildlife management classically focused on populations, habitat, and people (Giles 1978).

Behavior of an animal must be considered in wildlife management or when dealing with conservation issues (Yahner and Mahan 1997, 2002). For instance, here are six examples where the role of animal behavior must be considered (Martin 1998, Blumstein and Fernández-Juricic 2004). (1) Assume we are interested in understanding the social relationships in the management of brown bears (Fig. 1.1) (*Ursus horribilis*) or African lions (*Panthera leo*). In this case, we need to appreciate the role of infanticide in these species (Bertram 1978). (2) If a certain species is involved with our captive propagation, such as the California condor (*Gymnogyps californianus*), we need to ensure that individuals released into the wild know their predators or do not associate humans with food. (3) Response of animals to humans need to be considered; for example, whether brown bear attacks on humans are on the rise (Herrero et al. 2005), or whether the presence of humans affects the behavior of animals (Cooper et al. 2008). (4) Habitat loss is a major factor affecting the endangered status of animals (Wilcox and Murphy 1985). If this is the case, provision of habitat plus the addition of key features (e.g., nest boxes) needs to be understood in relation to spacing patterns and the species in question. For instance, if a species has a territory of a given size, placing nest boxes too close to each other may result in fewer boxes being used by the nesting species in comparison to a species with a smaller territory. (5) What the difference between proximate and ultimate factors is

Fig. 1.1 The brown bear, which is also called the grizzly bear, is a large bear that ranges across most of North America (west of the Mississippi River) and in northern Europe and Asia. Infanticide, or the killing of young, has been widely studied in humans, but it also occurs in brown bear and other animals ranging from rotifers to other mammals

when a species is selecting a habitat or a portion of the habitat, which might be very expensive for a wildlife agency or the public to acquire. Does an animal select a mate or the factor of a given habitat, such as habitat area? (6) When the parental care is important for a given wildlife species. Is the species monogamous, when is parental care paramount to the success of the species, or are both parents required for successful rearing of young? If a wildlife manager knew this information, it could conceivably affect length and nature of a hunting season for that species.

My focus in this text is on wildlife, which I define as any animal that is either domesticated or nondomesticated, including pets and nonpets (Yahner 2000). Plants are eliminated from consideration.

1.2 Comparative Psychology Versus Ethology

The history of animal (hereafter referred to as wildlife) behavior over the years can be divided into two approaches: comparative psychology and ethology (Barash 1977). Comparative psychology is largely a discipline developed in North America and is mainly focused on learning by wildlife. In comparative psychology, operant conditioning or trial-and-error learning occurs when an animal performs a behavior to receive a reward; operant conditioning was developed by Thorndike (American), whereby an animal learns to behave and get a satisfying reward (Thorndike 1898). For example, a domestic cat (*Felis catus*) in a box learns to press a lever that opens a door to food. Behaviorism was developed in the 1950s by Skinner, who was one of the most famous American comparative psychologists. Skinner placed a hungry animal, for example, a Norway rat (*Rattus norvegicus*) in a "Skinner box" (Fig. 1.2). In this box, the animal manipulated a mechanism, such as a lever, to provide a food reward; learning is measured as the increased frequency of the behavior to receive

1.2 Comparative Psychology Versus Ethology

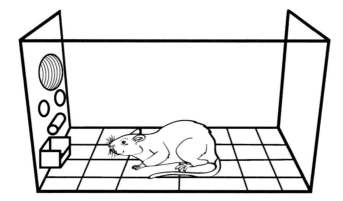

Fig. 1.2 An American psychologist, B.F. Skinner, invented an operant conditioning chamber, which measured the response of organisms (typically, rats) to a lever that provided food to the organism

the reward over time. According to this reinforcement of controlled behavior, basic learning processes are common to all wildlife species; thus, information obtained from Norway rats could be extrapolated to all species. Operant conditioning can be quantified as the time it takes an animal to learn the response, which becomes less as animal learns the response (the "law of effect").

Then, Pavlov, a Russian psychologist, introduced the concept of a conditioned reflex (Pavlov 1927). In this case, a domestic dog (*Canis familiaris*) salivates at the sight of food, with the sight of food signaling the presence of food. At this point, Pavlov would ring a bell before feeding the dog, and he found that dog salivated with the sound of a bell.

Today, operant conditioning (or behaviorism) and classical conditioning play an important role in wildlife conservation (Bauer 2005), as for instance, when dealing with nuisance wildlife. Physical (shots in the flank with rubber buckshot) and auditory (firecrackers) stimuli have been used to reduce feeding on garbage by nuisance black bear (*Ursus americanus*) in West Virginia (Weaver et al. 2004). Despite the use of operant conditioning and moving these bears 11 km from the nuisance site, all bears returned near or to the site and continued their nuisance behavior within 2 weeks. Although the use of operant conditioning in this example was unsuccessful, many successful examples of conditioning are in the literature to mitigate the impact of nuisance animals (Conover 1999). The wildlife profession must deal with nuisance wildlife because an estimated 75,000 people per year in USA are injured or become ill due to interactions with wildlife, via bites, vehicular collisions, etc. (Conover et al. 1995). Canada geese (*Branta canadensis*) have become a problem in many urban or suburban areas. To reduce or disperse Canada geese populations, chemical stimuli have been used on food used by geese that either imparts a bad taste that is irritating (anthranilate) or sickens the animal (methiocarb). Interestingly, coexisting mute swans (*Cygnus olor*) fed individually or in small group, as true of mute swans, were not affected by these chemicals; so, living in small groups does not allow the use of conditioning of food because aversion of treated food is not passed on to other individuals as readily.

In contrast to operant conditioning (or behaviorism) and classical conditioning, ethology is the evolutionary basis of behavior. Ethology was founded largely by Europeans and invokes natural selection as acting on behavioral traits that are inherited rather than traits that are learned. A classical ethological concept is the fixed action pattern (FAP) (Eibel-Eibesfeldt 1970). A FAP often is constant, e.g., the courtship display of bird species, which is triggered by a sign stimulus or releaser, such as a given color pattern on a bird. Alternately, a FAP may occur when a graylag goose (*Anser anser*) retrieves an egg that rolls out of the nest.

In short, wildlife behavior is a grand subject that integrates genetics, development, endocrinology, neurobiology, learning, evolution, and sociality from the combined insights of biologists, anthropologists, mathematicians, psychologists, and other specialty scientists. However, wildlife behavior was not always neatly integrated, which led to the "nature versus nurture controversy" (Lehrman 1970). In the mid-nineteenth century, there was a heated debate over whether behavior was inherited or learned. In other words, these early European ethologists, such as Tinbergen, believed that learning as a subordinate to instinct and innate behavior. They felt that behavior was adaptive and thus evolved, whereas American psychologists, such as Skinner, firmly felt that behavior was learned. Moreover, ethologists studied wildlife behavior mainly in natural environments, whereas psychologists did studies in the laboratory, under more controlled conditions. Today, however, ethologists, psychologists, or wildlife behaviorists in general assert that behavior of wildlife is shaped by both environment and inheritance. Thus, the nature versus nurture controversy largely has been dispelled.

Tinbergen (1963) considered four questions that are vital today to any serious student of wildlife behavior. These are (1) what are the mechanisms that cause a certain behavior? (e.g., hormonal, genetic, learning, etc.), (2) how does a given behavior develop? (e.g., ontogeny, cultural transmission, etc.), (3) what is the survival value of a given behavior? and (4) how does a given behavior evolve?

Chapter 2
Genetics and Other Mechanisms Affecting Behavior

2.1 Introduction

Innate behavior, e.g., behavior that is genetically programmed, enables certain organisms, such as spiders to spin a web (some species in the order Araneae). Use of highly inbred white mice, which are actually are inbred strains of the wild house mice (*Mus musculus*), by comparative psychologists is important as models of human disease or to demonstrate the effects of genetic versus the environment on behavior. However, even if a specific gene or set of genes for a given behavior is present in an organism, that behavior may not necessarily be shown. For instance, male birds usually do not sing outside the breeding season, because certain genes may not be turned on by hormones or nerve activity (Gill 1990).

2.2 Diversity in Behavior Diversity

Patterns of diversity in behavior are often similar among genetically similar species (Brown 1973). For example, many ducks and geese (family Anatidae) feed and forage the same way. Pigeons (*Columba livia*) drink water by sucking it into their throats with beaks in water (like mammals); in contrast, other bird species pick up their heads to let water go into throat via gravity. Some new species have been discovered using behavior. Prairie dogs (genus *Cynomys*) and fireflies (family Lampyridae) have been segregated into different species on the differential behavior (vocalizations and patterns of flashing, respectively). Fighting versus retriever ability has been selected in dogs (e.g., pit bull versus Labrador retriever). Dinosaur teeth, based on fossil records, have been used to distinguish predatory versus prey ways-of-life.

2.3 Sociobiology and Behavioral Ecology

Perhaps the most important concept in the study of animal behavior has been the idea proposed by Charles Darwin that evolution through natural selection has provided the necessary framework for the development of traits (Goodenough et al. 2001). Sociobiology and behavioral ecology are the application of evolutionary biology and ecology to social behavior, which were developed in the 1960s onward (Wilson 1975; Barash 1977; Goodenough et al. 2001). According to these two disciplines, ecological factors better correlate to behavior than phylogenic or genetic factors. In my view, sociobiology is not segregated from behavioral ecology; both are the focus of what many behaviorists study in natural populations.

Most behaviorists concur that natural selection acts on individuals, families or kinship groups (ants, bees), or spatially defined populations (demes) (Brown 1973). Natural selection acting on the latter group, however, is questionable. Another concurrence among most behaviorists is that fitness is mediated by natural selection and is expressed by differential natality and mortality.

2.4 Social Organization or Social System

The totality of interactions within and among species can be viewed as the social organization or social system of a species (Crook et al. 1976). Each aspect of this organization can be affected by selection, either natural or human-induced factors in the landscape. Seven aspects of social organization may be recognized: social units (composition and dispersion patterns), food-acquisition behavior (foraging behavior), predator-defense behavior, mate-acquisition behavior (sexual behavior), parental-care behavior, spacing patterns (e.g., territoriality), and communication systems. Communication of various means mediates all of the previous aspects of the social organization or social system of a species.

2.5 Social Units

Animals are distributed in the available or preferred habitat. The spatial distribution or dispersion pattern is determined mainly by behavior towards others of the same species (conspecifics) or other species (prey, predators), or by resources in the environment [e.g., spring seeps or steams in water-dependent salamanders, as the northern spring salamander (*Gyrinophilus porphyriticus*)] (Shaffer 1995). Typically, a given dispersion pattern is characteristic of a species; for instance, some species live in colonies, such as the yellow-bellied marmot (*Marmota flaviventris*) of the western states (Armitage 1998), whereas others, such as the woodchuck or groundhog (*Marmota monax*) of the eastern states, are solitary (Davis 1967) (Fig. 2.1).

2.5 Social Units

Fig. 2.1 The woodchuck, or groundhog, is the only marmot in eastern North America. Unlike its western counterparts, the woodchuck is solitary; western marmots tend to be colonial

Spatial distribution or dispersion pattern can be of three basic types: random, clumped, or regular (sometimes termed uniform) (social behavior from fish to man text). A random pattern is rare in nature; it may occur in a few species, maybe in a few species, such as the Poli's stellate barnacle (*Chthamalus stellatus*) on shoreline rocks, which may be a random resource for attachment. A clumped pattern is common in many social animals, such as a colony of ants (Formicidae) or herd of wildebeest (*Connochaetes* spp.). A regular or uniform distribution is found in species that defend a space via aggression (e.g., of territorial), as found in the songbird called the common yellowthroat (*Geothlypis trichas*).

Compared to an aggregation, a social unit has (1) a complex system of communication, (2) a division of labor based on specialization, (3) cohesion or tendency for members to stay together, (4) degree of permanency of group members, (5) and tendency to be impermeable to nongroup members [e.g., members of a gray wolf (*Canis lupus*) pack may kill a lone wolf] (Eisenberg 1966).

An example of an aggregation may be several brown bears in close proximity while foraging in a stream full of salmon (family Salmonidae) in western states. Group feeding, as with brown bears at a salmon run or with a flock of wintering birds, may increase the efficiency at which food is gathered (Morse 1970). For instance, wintering flocks of birds often contain several species, which later presumably pair and mate in the breeding season (Brown 1973; Rollfinke and Yahner 1991). In central Pennsylvania, over 50% of 123 wintering bird flocks contained four species, including downy woodpeckers (*Picoides pubescens*), black-capped chickadees (*Poecile atricapillus*), white-breasted nuthatches (*Sitta carolinensis*), and golden-crowned kinglets (*Regulus satrapa*) (Rollfinke and Yahner 1991). Of these flocks, there was a mean of 8.2 individuals and 3.8 species. Chickadees are often the most abundant flock members, perhaps because they have a distinct alarm system; as a result, other species may flock with them (Brown 1973). Hence, in harsh northern winters, flock formation may enhance (1) the ability to find food, and (2) decrease predation risk (more eyes) (Brown 1973; Bertram 1978). Some species may even form postbreeding flocks prior to migration to wintering areas (Hobson and Van Wilgenburg 2006). Some bird species in northern boreal forests, such as Tennessee (*Vermivora peregrina*) and yellow-rumped warblers (*Dendroica coronata*), form flocks of about 20 individuals as early as late June, and the flock size continues to increase to over 60 prior to southerly migration. Probably, the earliest birds to become members of a postbreeding flock were nonbreeders or "floaters" in the population.

Similarly, why do turkey (*Cathartes aura*) and black vultures (*Coragyps atratus*) form aggregations by roosting communally during winter in Pennsylvania and Maryland (Thompson et al. 1990). Winter roosts have 13–200 black vultures and 43–429 turkey vultures. Vultures selected winter roost with coniferous trees to provide favorable microclimate. Because black vultures cannot smell well, if at all, and turkey vultures have a keen sense of smell, the information-center hypothesis has been given for the use of the same roost by both species of vultures (Wright et al. 1986). Turkey vultures find carrion and other food items in the morning, and this food information then is transferred to the more aggressive black vulture who capitalizes on the food.

Bald eagles (*Haliaeetus leucocephalus*) roost communally in winter in Oregon and California (Keister et al. 1987). I probably would characterize this roost as an aggregation rather than a social unit. Bald eagles commonly exhibit winter roosts throughout their range. Sometimes as many as 500 eagles may roost overnight within 0.25–24 km of major daytime major concentrations of food, such as fish and waterfowl. In these roosts, subadults tend to roost closer to food resources, suggesting that, compared to adult birds, subadults were less efficient at foraging than adults.

In contrast to vultures or eagles, semipalmated sandpipers (*Calidris pusilla*), which are found in subarctic and Arctic regions of North America form nesting colonies of 2–4 pairs/ha (Jehl 2006). The semipalmated sandpiper is the only one of three sandpipers ($n=31$ species) that is known to form nesting colonies. This species is monogamous and highly territorial, which facilitates early nesting and identification of individuals, thereby making these pairs a network of social units rather than an aggregation.

A given social unit (or aggregation) and its dispersion pattern can be determined by censuses, aerial photos, quadrant sampling, etc. For example, by plotting burrow systems used by marked eastern chipmunks (*Tamias striatus*) in 12 0.3-ha quadrants located in optimal habitat was used to determine a regular distribution of solitary animals in the habitat (Yahner 1978a). Mean variance of distances between neighboring burrow systems was compared statistically to a value of 1.0.

Social units and dispersion patterns can change temporally. For instance, in white-tailed deer (*Odocoileus virginianus*), nonwintering units in the Welder Refuge of Texas typically consist of a matriarch group of a female and her young from 1 to 2 generations (Hirth 1977); rage mean size of social units (matriarch group) exceeded four animals, perhaps because animal in the social unit (more eyes) could detect danger in the landscape. On the other hand, the matriarch group at the George Reserve in Michigan was typically less than three individuals and did not include the young of the oldest generation. The Reserve had small open areas (less than 25 ha) within the landscape, with a total amount of open areas = 22%. In winter, on the other hand, deer may form huge amorphous herds (probably more appropriately termed, aggregations) of both sexes and many age groups as occurred once at Gettysburg National Military Park (R.H. Yahner, personal observation).

2.6 Matriarchal Social Units

What is the significance of a matriarchal social unit to animals? Much needs to be learned about this phenomenon, particularly because wild populations often have been managed by culling or hunting older, larger individuals, with the assumption that older animals play little role in the success of a social unit (Fox 2003). In African elephants (*Loxodonta africana*), overall reproductive success of a social unit has been found to increase with older females as matriarchs for three reasons: these matriarchs have improved knowledge of social calls for protection (e.g., bunching calls) of food source locations and possess better parenting skills than younger females in the social unit.

2.7 Plasticity in Social Unit Size

Some species presumably have plasticity or flexibility in size of a social unit. For example, the coyote (*Canis latrans*) has a basic social unit of a mated pair [such as the red fox (*Vulpes vulpes*)], but the size of the social unit in coyotes can vary with habitat. In forest areas of the eastern deciduous forest, only about 33% of coyotes are pack forming (3–5 individuals), 33% are pairs, and 33% are solitary (Todd et al. 1981). In the eastern forest, major prey of coyotes is rodents (order Rodentia) and rabbits (order Lagomorpha). But in open areas, as in the plains of Oklahoma, coyotes often form packs of greater than five individuals (Stout 1982). Major prey item of coyote packs in the open plains are often large (such as adult white-tailed deer). Hence, prey size may dictate the size of social unit.

Large mammalian predators have two strategies to be an effective predator on large prey. Large cats, such as the mountain lion (*Puma concolor*), are large in body size and hunts solitarily on large prey. Members of the dog family hunt large prey by increasing the number of members in the hunting party (social unit). Red fox always form a pair; as a consequence, red fox feed on small food items. Larger pack size of gray wolves allows cooperative hunting. In Alaska, pack size in gray wolves is usually 5–8 animals (Thurber and Peterson 1991; JM 74:881). In addition to increased pack size in coyotes to enable hunting of larger prey, two other factors for increased size of social units in the open plains may be (1) cooperative defense of carrion (presumably against other coyote packs) and (2) delayed dispersal of young (winter in coyotes versus fall in fox) (Stout 1982).

Why then do the African lion (*Panthera leo*) form social units (prides) rather than hunting solitarily like other large cats? Unlike large dogs, e.g., gray wolf, prey capture seemingly is not related to prey size in the African lion. A solitary lion is as efficient in prey capture as pride of 5–6 females, if prey is abundant. Larger prey size in African lions is apparently related instead to defense of cubs against infanticidal males (Bertram 1978). In addition, larger pride size in African lions also aids in defense of territories against other prides of lions or clans of spotted hyenas (*Crocuta crocuta*).

2.8 What Do Animals Learn in a Social Unit?

What do animal learn in a social unit (Goodenough et al. 2001)? Obviously, animals need to learn a number of things to maximize their fitness—what foods to eat, how to find these food, what habitats to use or choose (if given a choice), territorial boundaries, the identity of neighbors versus intruders, with whom to mate, familial or social unit relationships (most of whom are probably kin), or how to be or interpret aggression. Can animals learn how to choose these things? This becomes an important question when trying to determine the impacts of land-management practices on wildlife (McDonough and Loughry 2005).

For example, at Tall Timbers Research Station in Florida, the impacts of prescribed burning and timber removal on the nine-banded armadillo (*Dasypus novemcinctus*), which is the only armadillo species in the USA, did not affect individual movements away from the burned area (McDonough and Loughry 2005). Instead, burning reduced body weights and the number of females lactating, thereby affecting reproductive fitness in this population of armadillos.

2.9 Cultural Transmission of Learning

Is protecting cultural diversity in wildlife as important as protecting genetic or species diversity (Fox 2003)? Like humans, wildlife can pass on cultural traits form one generation to the next.

Cultural transmission of behavior occurs when an animal learns the behavior from others by a process known as social learning (Goodenough et al. 2001). Probably one of the best-known examples has occurred in the great tit (*Parus major*) of Britain (Fisher and Hinde 1949). This bird learned to open milk bottles and drink cream from them; this behavior then spread among great tits throughout the British Isles. In order for there to be cultural transmission of behavior, an animal probably must be preadapted to this behavior; the origin of opening milk bottles may have been bark tearing on birch (*Betula* sp.) trees for use of birch bark in nests; bark tearing may have been a preadaptation for opening milk bottles. Another example of cultural evolution can be found in cetaceans, such as sperm whales (*Physeter macrocephalus*), that live in matriarch units of up to a dozen females, whereas males of the species tend to be solitary (Vaughan et al. 2000).

Sperm whales can dive as deep as 1,000 m for up to 70 min in search of food. The matriarch unit of sperm whales uses a form of auditory communication, termed echolocation, to capture prey, to stay together as a social unit, and to avoid predators (e.g., humans). The population of sperm whales of the South Pacific can be subdivided into five clans, each with their own songs (dialects), movement patterns, and habitat use. A dialect can be defined as differences in vocalizations in one population to the next (Brown 1973). The adaptive value of cultural transmission of dialects in the sperm whale is that adult females conceivably imprint on dialects of their neighbors to serve as a reproductive isolating mechanism (Fox 2003). Hence, when conserving the sperm whale, what works as a conservation measure for a segment of the population may not work for another—one clan may use vocalizations or behavior in different ways to avoid whaling activities by humans. For instance, one clan of sperm whales may feed close to shore, whereas another may feed away from shores. If a given weather phenomenon, e.g., El Niño, affects food patterns, one clan may be more susceptible to whaling than another. Because culture is vital to ways to making a living and survival in humans, we are beginning to appreciate that culture diversity, and associated behavior in wildlife need to be considered in conservation.

Tool use in common chimpanzees (*Pan troglodytes*) has been seen most often in populations found in west Africa than those from east Africa. With dexterous hands and use of tools to acquire foods and for social displays in this species, the common chimpanzee is predisposed to this behavior via cultural transmission.

Scavengers often use cultural transmission of behavior to find food (Goodenough et al. 2001). As mentioned earlier, black vultures capitalize on the foraging efficiency by turkey vultures, which have a good sense of smell at communal winter roosts. This is a theory known as the information-center hypothesis, whereby animals learn about the location of food by interacting with others. This type of information transfer differs from individual learning because what an animal learns via individual learning stops when that animal dies. Observational learning, or cultural transmission of behavior, may increase the rate at which food is found (Morse 1970); for instance, sandwich terns (*Sterna sandvicensis*) take food from adults until midwinter, at which time they no longer take food from adults (Fig. 2.2).

By observing the foraging behavior of adults, young sandwich terns learn how to forage efficiently. As observers of this behavior, we can surmise that young are less dependent on parents.

2.10 Ultimate Versus Proximate Factors in Wildlife Behavior

When referring to ultimate factors, typically are asking how does evolution via natural selection shape behavior. Proximate factors are not evolutionary in nature. For instance, red plumage in house finches (*Carpodacus mexicanus*) are the result

Fig. 2.2 The sandwich tern is a seabird, in which males plunge for fish and offer the fish caught to females as part of the courtship. Young sandwich terns learn food items from both parents via observational learning

of carotenoid color pigments that are consumed in food (e.g., hinged to diet) (Hill 1993), but not clear what foods provide these pigments. Female house finches are not reddish; thus, differences in color between the sexes seem to be caused by differences in dietary intakes of carotenoid pigments. Females do not benefit by being colorful, so they focus on maximizing caloric intake per unit time; therefore, for both males and females, food is the proximate factor. Brightly colored males were significantly more likely to get a mate than lightly colored males, so being very red is the ultimate factor for male house finches. Because there is a positive correlation between male coloration and his parental care, whereby redder males are better mates by feeding females and chicks at a higher rate than less red males, survival of a chick and fitness of males are enhanced.

2.11 Hormones and Proximate Factors

The endocrine system consists of ductless glands that secrete hormones, which are molecular messengers secreted directly into the bloodstream (Goodenough et al. 2001). Hormones can affect behavior, as in adult male mammals where hormones initiate and maintain sex characteristics (as in the antlers of white-tailed deer).

Hormones in males are especially at high levels in the blood before male–male encounters. Testosterone, which is a male hormone, acts as a proximate factor affecting fighting ability and "play fighting" in younger animals. Proximate factors address questions such as "How is it that…" "What is it that…" e.g., How does a bird learn this? Ultimate factors get at the whys—e.g., "Why is it adaptive for a bird to learn this, e.g., a male red-winged blackbird to learn breeding habitat?"

2.12 The Nervous System, Biochemistry, and Behavior

The nervous system provides an electrical impulse via nerve cells, which consist of dendrites at its ends and an axon (long body) (Goodenough et al. 2001). Nerve cells provide mechanisms for instantaneous responses in wildlife. These instantaneous responses are necessary for all behaviors from flight in migration to feeding on a preferred food item.

In the vertebrate brain, a structure, called the hippocampus is found just inside the temporal lobes of the brain (Goodenough et al. 2001). The hippocampus (derived from Greek) has a curved shape resembling a seahorse. Phylogenetically, the hippocampus is one of the oldest parts of the brain; in humans suspected of having Alzheimer's disease, the hippocampus is one of the first regions to suffer damage, causing memory loss and disorientation. In vertebrates, memory and navigation is based on the hippocampus.

In corvids, the hippocampus of the brain is important (Krebs et al. 1989), by enabling these birds the ability to retrieve food. The large hippocampal volume in the brain helps these birds find 6,000–11,000 seeds stored after a period of 9 months. In gray squirrels (*Sciurus carolinensis*), the hippocampus is vital in finding the spatial location of acorns stored in the forest floor during autumn.

In invertebrates, the hippocampus does not exist. Rather, insects, such as honeybees (*Apis mellifera*) have a cluster of small neuron cells called mushroom bodies (corpora pedunculata) (Withers et al. 1993). The bodies enable a honeybee to learn foraging locations. Foraging bees have mushroom bodies that are about 15% larger than nonforagers.

A good example of how a biochemical factor affects behavior is with the ability of some species of animals, including fish, amphibians, reptiles, birds, and mammals to see ultraviolet (UV) light (Goodenough et al. 2001). For instance, in zebra finches (*Taeniopygia guttata*), a single change in an amino acid (C84S) can cause change in UV perception; the loss of this amino acid resulted in violet perception only!

Chapter 3
Mate-acquisition and Parental-Care Systems

3.1 Courtship and Mating Systems

Courtship in wildlife has always been fascinating to study (Brown 1973). Some questions that can be posed are: Why are some species monogamous and others promiscuous or polygamous? Why do males fight over females in some species but not in others? Why is courtship so elaborate in some species but not in others? What factors influence mate choice and who does the choosing (males or females)?

Eisenberg (1966) classified mating into four types (1) brief pairing, in which a pair temporarily comes together to mate, then each returns to an isolated state, e.g., eastern chipmunks; (2) tending bond or consortship, a pair breaks off temporarily from a social group to mates, e.g., American bison (*Bison bison*) herds, (3) pair bonds, whereby a male and a female remain together after copulation as a unit, e.g., coyotes, or (4) harem formation, in which a male associates with and actively defends a group of females against other adult males, e.g., white-tailed deer. With the exception of pair-bond formation, each of these types may be mating systems that are either polygamous or promiscuous, because the male may mate with a number of different females.

3.2 Monogamy

Monogamy is a mating system in which one member mates with one member of the opposite sex (Brown 1973). This type of mating system is rare in mammals, but commonplace in birds (90% of species); monogamy is not preferred by males of most vertebrates, so it is probably a secondarily derived condition (Barash 1977). Monogamy is expected in species when both sexes are needed to raise the young; because lactation is solely done by mammalian females only in raising young (Vaughan et al. 2000), a type of mating system is related to the type of parental-care system. How parents

Fig. 3.1 The yellow warbler is very wide-spread, extending its range throughout North America and as far south as central Mexico

raise their young may be different in closely related species. For example, in the yellow warbler (*Dendroica petechia*), a male regularly feeds its mate while his mate incubates the eggs (Morse 1980). But in the related black-throated green warbler (*D. virens*), however, has males that do not feed young, but spend time defending territories (Fig. 3.1).

In some cases, it may be too difficult to guard more than one mate from other males. This is shown in Kirk's dik-dik (*Madoqua kirkii*), which is a small antelope from Africa (Brotherton and Rhodes 1996). Another example is with the coral-reef fish (*Valenciennea strigata*), where the male cares for the eggs, and the female guards the male; hence, mate guarding may lead to monogamy (Reavis 1997).

Monogamy also may be found when a male has difficulty monopolizing more than one female simultaneously. An example of this is shown in termites (Wilson 1971). During a nuptial flight, an airborne male will attempt to mate with a female. A female will not mate until she is in a safe burrow. Thus, a male mates with a single female that he manages to defend from other males.

Occasionally, monogamy is separated into perennial and seasonal monogamy (Eisenberg 1966). If a species mates with the same individual for life, perennial monogamy is invoked, as in Canada geese (*Branta canadensis*) or in swan geese (*Anser cygnoides*). In the swan goose, the adult male is very vigilant during the first

4 weeks of brood rearing, whereas the female spends considerable time feeding (Randler 2007). After this approximate time period, the brood reverses the parental role by exhibiting decreased feeding and increased vigilance.

Seasonal monogamy occurs often in migratory songbirds (Gill 1990). In this type of monogamy, a pair stays together during the mating season, but is separate during the nonbreeding season. Presumably, it is hard for migratory songbirds, which may migrate considerable distances and are short-lived, exhibit perennial monogamy.

Infidelity is presumably common among wildlife, regardless of the mating system (Morell 1998). For instance, in 180 monogamous bird species, only 10% were found to be genetically monogamous.

3.3 Polygamy

Polygamy is when an individual has two or more mates, none of which is mated to another individual (Eisenberg 1966). Polygamy may be divided into two types. Polygyny is when one male mates with two or more females, and polyandry where one female mates with two or more males. Polygyny is very common in mammals but is found only in 2% of bird species; it is the most common mating system among vertebrates, including humans (Goodenough et al. 2001).

Fur seals (family Otariidae) are probably the most conspicuous polygamous mammal, with males showing intrasexual competition via dominance behavior mammal (Brown 1973) (Fig. 3.2). Fur seals are found along the northern Pacific coast.

When the breeding season approaches in spring, male fur seals come to shore and establish territories on shore; males defend harems, which usually average 16 females. Mating can span an average of 31 days; in males, sexual maturity can be delayed (as true of many polygamous species) until 7 years but only 3–4 years in females.

Female fur seals come ashore in June–July to mate and have pups (Brown 1973). Most (90%) older females (6–10-year-old) that come ashore are already pregnant; these females give birth soon after coming ashore (within 0–3 days), and then become into breeding condition (come into estrus) about 4–7 days after giving birth.

Preferred male territories of fur seals on the shore are held by the strongest males and are near the edge of the water (unlike a lek, see below) (Brown 1973). Males defend these territories for up to 2 months without feeding; fights can be fierce and many males may die, with mortality rate in males is threefold that of females, "rounds"-up females into a harem, which is defended via threats and fights with other males.

Body size and, hence, sexual dimorphism may occur in harem-forming mammals (Brown 1973). The polygamous species of red deer of Europe (which is the same species as the elk, *Cervus canadensis*, of North America) has males that "round"-up females into harems in the autumn. These harems are defended from other males via threats and fights. Like other harem-forming species of deer (as with the white-tailed deer), males leave the harem after the autumnal breeding season, whereas females form matriarchal groups. In contrast to red deer, both adult male and adult female

Fig. 3.2 Fur seals consist of several species, with the northern fur seals occurring in the North Pacific and most other species occurring south of the equator. Typically, fur seals form assemblages each year during summer months to give birth and breed. All species are polygynous (a dominant male reproducing with more than one female)

roe deer (*Capreolus capreolus*) are similar in body size. Roe deer usually do not form harems; intrasexual fighting in adult male roe deer is less intense than in red deer, and antlers in male roe deer are quite small. In roe deer, males and females occur as monogamous pairs.

A third type of polygyny recognized in wildlife is lek behavior. In lek polygyny, males defend territories that are traditional display sites. Females visit these leks, select a mate, copulate, and leave. Leks occur in 16 families of birds from grouse to hummingbird families (Welty 1973); in these species, males have striking sexual dimorphism in both plumage and behavior. A well-known example is the greater sage-grouse (*Centrocercus urophasianus*), which uses visual displays to attract potential mates. In this bird, leks may be 1,000 m × 200 m and contain about 400 males. However, only about 10% of the males in the center of the lek perform more that 75% of the copulations, with the youngest males on the periphery of the lek (after Gill 1990). Translocations of the greater sage-grouse, which is a declining species in UT, have shown that 100% of the surviving females are found in leks (Baxter et al. 2008).

Mammals, such as the male Uganda kob (*Uganda kob*), also use visual displays to attract mates on leks (Vaughan et al. 2000). These leks may consist of 30–40 males, each spaced 15–30 m apart. The hammer-headed fruit bat (*Hypsignathus monstrous*) of Ethiopia establishes traditional leks along streams (Bradbury 1977);

3.3 Polygamy

instead of using visual display to attract females, males use vocalizations to attract mates and establish dominance. As with other polygynous species, males are sexually dimorphic from females. Males are larger, plus they have an enlarged rostrum, larynx, and head.

In North America, red (*Lasiurus borealis*) and hoary bats (*L. cinereus*) frequently are killed by wind turbines designed for energy production (Cryan 2008). These bats are solitary and migratory, but they may be gregarious in late summer and fall. There is some speculation that these species may possibly form leks, and lekking behavior near a turbine may represent a tall structure (typically a tree) to visually look for bark under which to roost singly. Mating in both species occurs in the fall; males are not larger than females but may be of a slightly different color. Whether or not these bats use visual or auditory cues (if lek forming) to attract mates or eventually learn to avoid turbines.

Extra-pair matings, as found in polygamy, have benefits and cost to males. If a male mates with more than one female, he may spend time and energy in search of other female, thereby passing on his genes to several females. However, while a male is searching for other mates, the female may mate with other males. Benefits accrued to females in mating with several males may be that because extra males may help defend a territory from predators. Also, a female may be assured that ample sperm is available to ensure fertilization of all eggs.

Benefits of leks to males include an increase in the range of signals to females, rather than signals given by a sole male, males may require specific display habitats that are limited and patchily distributed, more eyes provide protection from predators, leks may serve as information centers about good foraging areas, and less successful males are better off being in association with successful males (e.g., summarized in Goodenough et al. 2001). Benefits of leks to females may be as many as three: leks provide many males in one location to select from and with whom to mate, and leks may reduce predation risks or competition with males for resources because males are congregated. Obvious costs of polygyny to females, including that of lekking behavior, is that there is no help from males in rearing of the young, plus females must share resources with other females. However, there must be a benefit to females in order for polygyny to evolve.

Recently, harvest effects have been shown to have some effect on polygyny (Whitman et al. 2007). In Tanzania, trophy hunting of African lions is limited to males, and less than 2% of the 15,000–20,000 lions are harvested. But to Tanzania, hunting (harvest) economics are lucrative. Each client pays $850/day to harvest a lion, with a trophy fee of $2,000; for 21 days on a safari, which amounts to an estimated $30,000–50,000 per person. Perhaps as a function of the harvest of male African lions, pride (group) size of females can be reduced to 0 if males greater than 3-year-old are harvested, but pride size is reduced only about 30% when males greater than 6-years-old are harvested. Therefore, hunters of African lions need to recognize older males in hunted populations, with the nose more pigmented in older animals. Furthermore, hunters can assist the local people by removing problem animals near high-human areas for sustainable management. Local people can be helped if nuisance or problem lions, e.g., those feeding on domestic animals are removed.

Compared to polygyny, polyandry is much rarer in the wild (Eisenberg 1966). In the bird called the northern or American jaçana (*Jacana spinosa*), polyandry occurs (Graul et al. 1977). Females are colorful, dominant to males in courtship, and territorial. After mating and being courted by the female jaçana, the male incubates the eggs. The male is the only sex with incubation patch, which is a highly vascularized bare patch of skin on the belly. Once young hatch, a male protects the young. Each female may have as many as four males within her territory. She is dominant in territorial defense and in courtship; thus, a number of females may be excluded from breeding. At least one-half of the nests are lost to predators. Therefore, females invest as much time into laying multiple clutches, and sex-role reversal may be adaptive.

The spotted sandpiper (*Actitis macularia*) is also an example of a polyandrous bird (Maxson and Oring 1980). This bird is a widespread shorebird in which the female is larger than the male and is territorial. The female spotted sandpiper may lay separate clutches with one or more males; these males defend these all-purpose territories from other females. In a sense, these females exhibit the strategy of bet hedging by mating with other males as they show up on her territory. Perhaps this may occur as a result of high rates of nest predation.

In humans, fraternal polyandry is a form of polyandry in which two or more brothers share one wife or more. Fraternal polyandry is accepted in certain areas of Tibet (Levine 1988). In general, however, polyandry in human or animal societies is expected to evolve less often than polygyny because sperm from one male is often sufficient to fertilize all eggs (Goodenough et al. 2001). Thus, seldom it is advantageous for a female to mate with greater than one male; from a male perspective, there is little value in polyandry.

3.4 Promiscuity and Other Mating Systems

Promiscuity does not involve pair-bond formation (Eisenberg 1966) and occurs when males and females mate with one to many of the opposite sex; no one has exclusive rights over individuals of the opposite sex with promiscuity. Promiscuity is common in mammalian species, but it is found in only 6% of bird species. Males that are promiscuous often have large home ranges and are very mobile during the breeding season. An example of a promiscuous mammal is the eastern chipmunk (Yahner 1978b).

Mating chases in the promiscuous eastern chipmunks occur twice per year in southerly latitudes of its range (Yahner and Svendsen 1978). Estrus usually occurs in late winter (February–March) and summer (July). During estrus, adult males routinely inspect burrow systems of adult females at sunrise several days before mating chase. On the day of estrus, the adult female is active in the morning, and 2–8 males usually chase the female (Yahner 1978b). Males in the chase are very aggressive toward each other; they presumably follow the female via vision and olfaction. Copulation takes

3.4 Promiscuity and Other Mating Systems

place in a protected area (burrow, under tree roots); there is some circumstantial evidence of multiple insemination (several males transfer sperm into a female).

Hermaphroditism occurs when the same individual has both functioning male and female sex organs and is capable of self-fertilization; well-known examples include the oligochaetes (order Oligochaeta), which contain earthworms (Stewart 2004). Hermaphroditism is most common in invertebrates, but it does occur in some vertebrates, as in the fish called the mangrove killifish (*Rivulus marmoratus*).

Another term related to a mating system is protogynous hermaphroditism. A protogynous hermaphrodite is an animal that begins its life cycle as a female, but as the animal ages and based on internal or external triggers, it shifts sex to become a male animal. Some fish, e.g., lyretail coralfish (*Pseudanthias squamipinnis*) undergo this metamorphosis (Lieske and Myers 2004). In the coralfish, the male maintains a harem of 5–10 females. If the male dies, one of the females will undergo sex reversal and take the place of the missing male.

In reserves in Ghana, West Africa, recent research suggests that in large mammals, it is better to live in big harems than as pairs or in small harems (Brashares 2003). Over 50% of the mammal species in Ghana have become locally extinct ($n=78$ populations, including 9 primate species, 24 ungulate species, and 8 carnivore species) since the 1970s because of overhunting and habitat fragmentation or loss. Hence, monogamous species may be more susceptible to extinction. Several duiker species (subfamily Cephalophinae), which are small monogamous antelopes died out locally after only 10 years since the reserve establishment. In contrast, the African buffalo (*Syncerus caffer*) is still surviving on all of the reserves, and it occurs in harems of about 15 females. Similarly, several colobus (*Colobus* spp.) monkey species, which have few mates, died out over an average of 18 years after the reserve establishment, whereas green monkeys (*Chlorocebus* spp.), which have harems, are still present on the reserves. The link between extirpation of a population and harem size is unclear, but it may be related to hunters selecting males over females (because of greater size, presence of horns in males, etc.). Harem-forming species would have excess of males; also, perhaps smaller groups less likely to detect approaching predators, e.g., hunters. A bottom line may be that conservation efforts may be more difficult for monogamous or smaller groups than with species living in larger groups.

In the vast majority of internally fertilized species, sperm is the only contribution by the male; hence there seldom are sex-role reversals (Goodenough et al. 2001). Male, therefore, are not a limiting resource from a sexual-selection perspective, and males compete among themselves, with females doing the selection of a mate (see Sect. 3.6).

There are only a few known exceptions to internally fertilized species, with only males (and not females) caring for the young, that being the giant water bug (*Abedus herberti*) and the American or northern jaçana (Goodenough et al. 2001), which was discussed earlier as a polyandrous species. In the giant water bug, a female lays eggs on the back of a male; the male then takes care of young after this oviposition by aerating the eggs via rocking motion near the water–air interface. Thus, a limiting resource for the giant water bug is the space on the back of a male.

3.5 Mating and Mechanisms of Mating Interference

Males of some wildlife species are suspected of or are capable of multiple ejaculations or frequent copulations to make a given female less available to other males (Yahner 1978b; Hogg 1984; Jia et al. 2001; Toner and Adler 2004). However, males guarding females before or subsequent to mating is known in many wildlife species, thereby perhaps ensuring that the female stores sperm from only the mated male (Tsubaki et al. 1994).

If males are not large to compete for females or guard females as mates, male–male interference may become a factor. Probably the most extreme way of interfering with the reproduction of a rival male is to kill the young of the rival, which is termed infanticide (Hausfater and Hrdy 1984). Infanticide was found in many past human societies. Infanticide is known in many species of wildlife from rotifers (phylum Rotifera) to mammals, such as African lions and brown bears, but infanticide is also reported in some birds, such as tree swallows (*Tachycineta bicolor*).

Perhaps less extreme methods of sexual interference may involve spermatophores, thereby reducing the reproductive effort of rival males is shown in amphibians or the use of repellents or copulatory plugs. A spermatophore is a capsule or mass that is created by males of various animals and contains sperm. It may be transferred to the sexual organ of the female (cloaca) in its entirety during reproduction. Alternatively, the spermatophore may provide nourishment to the female. For example, salamanders, such as the red-spotted newt (*Notophthalmus viridescens*) (Massey 1988), have males that interfere with spermatophore transfer by (1) inserting himself between male and female just after the male dismounts the female, (2) exhibiting female behavior, causing the courting male to "waste" spermatophore, or (3) interfering with amplexus (Zug et al. 2001; in amphibians, male grasps female from behind to mate), causing the male that is mating to waste time and energy.

In the parasitic wasp (*Cotesia rubecula*), a recently mated male will mimic a female long enough for the mated female to become unreceptive (Field and Keller 1993). By mimicking a female, this postcopulatory guarding tactic increases reproductive success of the mated male. Male moths and butterflies produce two of sperms: eupyrene sperm, a viable sperm containing genetic material from the male to fertilize eggs, and apyrene sperm, which acts as the nonfertilizing sperm. Like kamikaze sperm, apyrene sperm can displace or deactivate the sperm of a rival male and perhaps act as a plug in the uterus of a female (Dumser 1980).

3.6 Mate Choice

Does the female choose a male as a mate or is it vice versa? Instead, does a female choose a male on the basis of the resources held by a male? For instance, males seem to be selected on the basis of their sale for salesmanship, whereas females selected for sales resistance (perhaps this is why males typically do most or all of the courtship, see Chapter 4; a female invests much more energy in

3.6 Mate Choice

developing an egg). Mate choice remains a difficult, if not unanswered, question that wildlife behaviorists have tried to answer. The male bullfrog (*Rana catesbeiana*) defends a space (territory) along lakes and ponds, with adequate vegetation for cover, from other males (Shaffer 1995). Apparently, a female selects the male based on the quality of his territory rather than on the quality of the male per se.

Choice of a mate, or more appropriately, which sex does the choosing may be related to the mode of fertilization (Goodenough et al. 2001). If fertilization is internal, then choice may by that of the female; but if fertilization is external, then it may be the choice of the male for at least three reasons. The first reason is called the certainty of paternity hypothesis. That is, if males genetically are related to the young, certainty of maternity guarantees that 50% of the genes of the female are present in each offspring; but males cannot be certain of paternity, if there a female is fertilized internally and insemination by multiple males is possible. Choice of a mate may be related to the gamete-order hypothesis, whereby parental care may be a function of differences in a male or female parent to desert offspring. In other words, whichever parent releases gametes last needs to raise the young. If internal fertilization occurs, the female is last to release her gametes; the reverse is true if here a species has external fertilization. Finally, a third reason for mate choice may be attributed to the association hypothesis. A female with internal fertilization would be near eggs or young when the young are born, so she cares for young; but if the male is territorial and there is external fertilization, he will be in association with the young and will exhibit paternal behavior. At least in most species of fishes and amphibians, the association hypothesis is believed to be the most viable hypothesis; these two taxonomic groups show both male and female care, and both internal and external fertilization.

Mate choice is not meant to be a term to imply that an animal makes a conscious or rational choice among potential mates, but rather it is an anthropomorphic way of saying that members of the opposite sex mate selectively. As indicated earlier, a female may select a male on the basis of the territory provided to the female by the male or valuable resources, like nest sites and food, contained within the territory. A female pronghorn (*Antilocapra americana*), for example, chooses a male with a territory that contains abundant forage (Kitchen 1984).

Females may select males on the basis of other criteria, such as nuptial gifts brought to her by the male, including spermatophores, seminal or glandular secretions, or prey. For instance, male fireflies (*Photinus* spp.) transfer a spermatophore to the female (Lewis et al. 2004) during copulation; part of the spermatophore fertilizes the female, with the remainder of the spermatophore providing a protein source for developing oocytes because she does not feed as an adult. In many arthropods, such as wolf spiders (family Lycosidae), males are typically viewed as a meal after mating (e.g., is cannibalized) by the much larger female (Persons and Uetz 2005).

Females may use physical or behavioral features to access mate quality. For instance, the red epaulet on the wing of a male red-winged blackbird (*Agelaius phoeniceus*) apparently is a good predictor (Eckert and Weatherhead 1987). Epaulet size was a reliable predictor of captive dominance rank in this bird (see Fig. 2.2).

Female choice of a given mate may not always occur. For example, in the swarming midge (*Tokunagayusurika akamusi*) males tend to swarm during mating, but males may mate with younger females as soon as they emerge, suggesting a lack of female choice of males (Otsuka et al. 1986). In captive breeding, males may be limited in their ability to select mates (Fox 2003). In fish hatcheries, mating may not be random. In captivity, we often choose who mates with whom, which may not depict the real world! One story comes from the Baltic Sea salmon (family Salmonidae) hatcheries, when in 1974, a mysterious blight killed 95% of the juvenile Atlantic salmon in the Baltic Sea. In the wild, females selected males with bright-red coloration brought out by carotenoids; these are antioxidant chemicals to enhance fighting ability and warding off pollutants and perhaps blight. Now, hatcheries treat fry with vitamin B1 to keep mortality low and allow red males to develop with nonrandom mating.

In the Chinese barking deer (*Muntiacus reevesi*) in semicaptivity, some male deer seemingly never had an interest in mating with receptive females (Yahner 1979). Only when a new male was put in with a female did she have young.

Chapter 4
Mating Systems and Parental Care

4.1 Courtship

In wildlife, there are three possible functions to courting the opposite sex, which is usually done by the male (Barash 1977). These functions serve as advertising signals, overcoming aggression, and achieving reproductive coordination between the courting male and a female or females, in the case of a polygamous or promiscuous species.

Courtship typically is prolonged in species with extensive parental care (Barash 1977). The first function of courtship, advertisement, can be divided into at least two components (1) to attract attention, e.g., here I am, I am a red-winged blackbird, and I am a male; and (2) to identity reproductive condition, e.g., I am ready to mate. A second possible function of courtship, to reduce aggression between prospective mates, would be especially important if the identity of an intruder is mistaken, as when sexes look alike, thereby resulting in possible injury, or even death, in one or both of the sexes. For instance, the male herring gull (*Larus argentatus*) has an eye ring that presumably is used by the female to distinguish the sex of her mate (Smith 1966). If the color of the eye ring of the male is altered, the female will not mate with the male.

4.2 Parental-Care Systems

Parents, regardless of wildlife species, usually are quite giving to their young as a mechanism to ensure that genes are passed on to the subsequent generation (Goodenough et al. 2001). Parents, therefore, have two decisions to make (1) how much of their resources should be allocated to reproduction versus growth and survival? and (2) how should resources be distributed among their offspring? Factors that may influence these two decision factors include the chance of remating during a lifetime. For instance, in

Fig. 4.1 The spotted hyena, or laughing hyena, occurs today only of the south of the Sahara, except in the basin of the Congo. Hyenas live in large matriarch groups, or clan, which may comprise 80 individuals. Hyenas are not scavengers, but actively hunt large mammals

some species with a short lifespan, mating may be a one-shot deal. Moreover, some parents must decide which young to feed, which may vary with size of parents, size of young, or order in which young are born. For example, in the giant egret or the great white egret (*Ardea alba*), the first young born may kill a younger nestling, termed sibling rivalry (Mock 1984). In both captive and wild populations of spotted hyena (*Crocuta crocuta*) (Fig. 4.1), sibling rivalry is common between twins born to a female (Wahaj et al. 2007).

Hence, overproduction of young will help ensure that at least one young reaches survival age when insufficient food is available.

Parent–offspring conflict was first discussed by Trivers (1974). This conflict is evolutionary in origin and arises from differences in parents tending to maximize their fitness (survival and reproductive output) while offspring are trying to maximize theirs. Offspring can increase their fitness by getting a greater share of resources and care from parents by competing with their siblings. These conflicts have led scientists, such as Lack (1947) to propose that wildlife (e.g., birds) has an optimal number of young per clutch or litter, which is determined by natural selection according to the maximum number of young that the parents can feed or nourish.

Why should phylogeny play a role in parental care? Biparental care is rare in amphibians, mammal, and reptiles (Vaughan et al. 2000; Zug et al. 2001), but it is common in birds (Gill 1990). In mammals, which have lactation as a universal feature in females, still have about 3% of males staying with females to rear the young. Perhaps there are two explanations for lactation never evolving in male mammals: there are limitations (1) imposed by food resource availability and not by lactation, and (2) on time and energy of a male, e.g., to guard its mate. On the other hand, parental duties, such as incubation, feeding, and guarding, are shared by both sexes, which requires both parents in raising the young.

4.3 Altruism and Parental Care

Why should an animal give aid to individuals of the same species, e.g., be altruistic, thereby increasing the reproductive fitness of another species at its own expense? Perhaps helpers, or altruistic animals, are more common than expected in nature, being reported in African lions and in many species of mammals and birds. In the African lion, a female has reportedly nursed young from other litters. Thus, helpers may be related to young and may be more common than expected in nature (Van Orsdol 1984).

The widespread nature of helpers suggests that there are advantages to having or being a helper, even when the helper is not related to the young. Helpers of young may assist in meeting the nutritive needs of the young, may give experience to the helper in raising young, may help the parents defend a high-quality resource, e.g., territory or nest site, and may be better than living alone as a means of finding food or detecting predators. A classic study of helpers in birds is with the Florida scrub jay (*Aphelocoma coerulescens*), where about one-half of the breeding pairs have helpers; some mated pairs may have as many as six helpers! Florida scrub jays apparently live in stable groups consisting of a breeding pair and young of 1 or more years (Woolfenden 1973). The pair selects the nest site, and the female incubates the eggs as she is fed by the male. When the young hatch, helpers bring food to them; these helpers also are vigilant at the nest and help mob predators (mobbing is discussed later). The number of young that subsequently left a nest with at least one helper averaged 2.2 compared to only 1.6 without helpers.

Another bird with nest helpers is the brown-headed nuthatch (*Sitta pusilla*), which is a species of concern occupying open pine (*Pinus* spp.) forests (Cox and Slater 2007). Some territories had five adult brown-headed nuthatches, with more than 70% of territories containing more than two adults. Totally, 15–30% of the territories consist of at least one helper, which is typically a second-year male helper. Unlike the Florida scrub jay, no difference was found in nest survival in the brown-headed nuthatch, having 4.4 young per nest with at least one helper and 4.2 young per nest without a helper. In this species, a helper may be beneficial if food is scarce or dead trees (snags) are lacking.

Another form of altruism is alarm calling; auditory communication, with the function as serving as an alarm, will be discussed later. An alarm call may benefit the hearers by alerting them to danger, but may risk the life of the caller by drawing attention to it. Mutual grooming in some primates and feather preening in some birds also have been considered a form of altruism.

4.4 Brood Parasitism and Parental Care

Brood parasitism occurs when a brood parasite, or a donor, does not raise the young; rather the duties of a parent are placed on another parent. Brood parasitism is best known and studied in birds, and it can be within the same species (intraspecific) or among species (interspecific). About 1–2% of extant (living) birds exhibit brood parasitism (Gill 1990). Brood parasitism may have arisen in tropics in response to high predation rates; also, the habit of using old or abandoned nests may have been precursor to parasitism. For instance, European starlings (*Sturnus vulgaris*) use old woodpecker (family Picidae) holes in trees as nest sites.

Intraspecific parasitism may be the first step in the evolution of obligate parasitism. Intraspecific brood parasitism is most common in waterfowl (e.g., ducks and geese). As in all instance of brood parasitism, individuals other than the genetic parent provide care for the young. Intraspecific brood parasitism is known in at least 53 species of birds, with 80% of these having precocial young (precocial versus altricial young will be discussed later). Intraspecific brood parasitism may involve one of the two strategies. In the first, eggs are "dumped" into nests of the same species (conspecifics); eggs then are incubated by the foster parent, and the foster parent then raises the young. Here, the donor may benefit, as in bar-headed geese (*Anser indicus*), which has 5–6% hatching success of donated eggs. Hence, this strategy is better than not producing eggs at all. However, recipients are affected negatively, with 29% hatching success of their own eggs, but the donor's eggs experience a 67% hatching success. In the second strategy, young of some waterfowl species mix young with those of conspecifics once hatched to form mixed broods, which are cared for by a foster parent. Therefore, this second strategy may be brood parasitism, in that donor benefits and recipient is negatively affected. But perhaps this second strategy is best called brood amalgamation (e.g., brood mixing).

Intraspecific brood parasitism is particularly common in waterfowl that nest in tree cavities or nest boxes, although cavity nesters comprise only 2% of the total bird species. In wood ducks (*Aix sponsa*), for instance, 95% of the nests in boxes were parasitized. Although some boxes were empty (suggesting that boxes were not limiting), some clutches had 30–40 eggs, compared to a normal clutch size of 10–12 eggs. Many heavily parasitized nests are abandoned. But over a 4-year period, the same female can be a host (75%) of the eggs of a conspecific, many were dumpers (54%), and some were both host and dumpers (23%). If no nest boxes were present and tree cavities are the only nest sites available, then parasitism is adaptive, as in historic times. Thus, it may be evolutionarily beneficial for wood ducks to use nest boxes, if available, today. Interspecific brood parasitism has been termed true parasitism, because the donor parents always benefit, but the recipient parents do not (Goodenough et al. 2001). Interspecific brood parasitism is prominent in birds known as honeyguides (family Indicatoridae) and in many of the cuckoo bird species of the Old World (subfamily Cuculinae) (Gill 1990).

There are several adaptations that a donor or its young has over the recipient and its young. One adaptation seen in the shiny cowbird (*Molothrus bonariensis*)

4.5 Brown-Headed Cowbird

Fig. 4.2 The shiny cowbird of South America, although it may breed in southern Florida. Like other cowbirds, the shiny cowbird is a brood parasite

(Fig. 4.2) is that the young cowbird may destroy eggs or kill nestlings of the host yellow-shouldered blackbird (*Agelaius xanthomus*) (Post and Wiley 1977). A second adaptation is that a donor may have eggs that closely resemble those of the primary host (Jensen 1980). A case in point is the parasitic didric cuckoo (*Chrysococcyx caprius*) and its host, the vitelline masked weaver (*Ploceus vitellinus*). Another adaptation is that the host, e.g., common cuckoo (*Cuculus canorus*) has eggs with similar patterns and color as that of its host, the great reed-warbler (*Acrocephalus arundinaceus*) (Payne 1977). In Asia, a large parasitic bird, known as the common koel (*Eudynamys scolopacea*), will distract host crows away from nest while the donor female slips in to lay eggs (Nicolai 1964). Often times, the parasitic brood parasite may lay more than one egg in the nest of the host, and, on average, the parasitic young hatch perhaps 2–4 days earlier than young of the host (Lack 1968; Payne 1973). Hence, this "head-start" allows hatchling parasites to grow faster by getting most of the parents (hosts) attention. Moreover, besides forgoing parental duty as a parasitic host, a bird may have more opportunity to breed. In the male yellow-rumped honeyguide (*Indicator xanthonotus*), which defends a feeding site of the honeycomb of a bee colony, he may mate with as many as 18 parasitic females per day, rather than helping raise a single brood (Cronin and Sherman 1976).

4.5 Brown-Headed Cowbird

The brown-headed cowbird is the only obligate parasite in North America, numbers in East have increased since 1900s with prairies going to agriculture (Yahner 2000). This may be responsible for the decline in Kirtland's warbler, an endangered songbird in Michigan (Probst and Weinrich 1993). From 1957 to 1971, 75% of nests of the Kirkland's warbler were parasitized by cowbirds, probably was the major factor causing a drop in the number of singing Kirtland's warblers from 502 to 201. Controls of cowbirds then were implemented. Unlike the common cuckoo, the brown-headed cowbird is not particularly specialized in parasitizing certain species. For instance, it probably parasitizes at least 50 different species. Eggs of the cowbird are of normal size and shell thickness, and colors do no match those of a given host. Cowbird young do not evict nest-mates of another species. But young cowbirds hatch earlier, and they are larger and stronger than young of hosts.

Some of the 50 species birds parasitized by brown-headed cowbirds are termed "acceptor" species (about 20%) because have not evolved defenses against cowbird [e.g., red-eyed vireo (*Vireo olivaceus*) or wood thrush (*Hylocichla mustelina*)]. These acceptor species readily "accept" eggs of cowbirds in nests. Because the cowbird is a relatively recent arrival in Eastern states, species may be acceptors as a result of an equilibrium lag hypothesis (Gates and Gysel 1978). Other forest bird species are known as "rejector" species (about 80%); in order to be a rejector, a bird species needs to recognize differences in eggs between those of cowbirds and those of their own. Rejector species either throw the cowbird eggs out of its nest [e.g., gray catbird (*Dumetella carolinensis*)] or buries them by placing another nest on top of the cowbird eggs [e.g., yellow warbler (*Dendroica petechia*)]. In the case of the long-term survival of the Kirtland's warbler; the warbler may never has time to develop "rejector" ability (Norris 2006).

Grassland songbirds are the fastest declining group of birds, but only <2% of nests of grassland songbirds were parasitized by cowbirds at a study site along the Kentucky–Tennessee border. This study suggested that species rejected cowbird eggs not because of an equilibrium lag hypothesis but because of evolutionary equilibrium hypothesis, which implied that even if ejection behavior had evolved in grassland birds [e.g., the Sprague's pipit (*Anthus spragueii*)], these grassland birds may accept cowbird eggs because ejection may be too costly. Grassland birds may exhibit recognition errors and reject its eggs rather than those of the cowbird, thereby causing ejection errors by damaging their own egg in the process of rejecting a cowbird egg.

An acceptor species, the wood thrush, has its nest parasitized when the incubating female leave the nest. In about 10% of these cases, a female cowbird will occupy a nest occupied by an incubating female wood thrush and aggressively attack the female to remove her from the nest.

The percentage of nests parasitized by the brown-headed cowbird can vary temporally and regionally (Hoover and Brittingham 1993; Bollinger and Linder 1994; Yahner and Ross 1995). In Illinois, 75% of the nests of wood thrush were parasitized; 5 years later on the same site, only 25% of the wood thrush nests was parasitized by cowbirds. In Kansas, 21% of the nests of 14 species were parasitized, compared to 48% of the nests of a variety of birds in Wisconsin. Because the cowbird is considered an edge species, brood parasitism by cowbirds is expected to be greater in species also considered as edge species (Yahner 2000). In Wisconsin, 65% of the total nests within 100 m of edge were parasitized compared to only 3% of the nests established 300 m or more from the edge. In Pennsylvania, no relationship was found between the number of nests near an edge and the percentage of nests parasitized (Giocomo 1998). Actually, the reverse occurred in Vermont, with only 7% of bird nests placed near edges being parasitized but 32% of the total nest in forest interiors being parasitized by cowbirds (Hahn and Hatfield 1995). There is some question whether brood parasitism occurs in other taxonomic groups, such as fish (Goodenough et al. 2001). In the mouth-brooding, predaceous cichlids (family Cichlidae) of Africa, females pick up their eggs soon after laying them and brood them in their mouth until the fry (young) are ready to emerge. If danger is in the area, the young escape into the mouth of mother. In these cases, broods from other cichlid species (e.g., plankton-feeding cichlid) have been found in the mouth of the predaceous cichlid. Whether this is foster parenting in cichlids as true of acceptor bird species and brood parasitism by birds is open to debate. In insects, such as cuckoo bees, brood parasitism occurs in a variety of lineages (subfamily Nomadinae) (Michener 2000). These bees lay eggs in nests of other bees, and when these eggs hatch, the larva of the cuckoo bee eat the larva of the host species.

4.6 Altricial and Precocial Young

Parental care is common in many vertebrates, including reptiles, birds, and mammals (Gill 1990; Vaughan et al. 2000; Zug et al. 2001). This phenomenon has been called the single most striking feature of postnatal growth. Altricial birds and mammals have several characteristics compared to precocial bird and mammals: they are (1) naked, e.g., featherless or hairless, (2) are blind or virtually blind, (3) tend to be immobile, (4) in birds, eggs of altricial are smaller than those of precocial species per unit size of adult (altricial eggs with 15–27% yolk; precocial eggs with 30–40% yolk), and (5) have a relatively small brain size. Examples of altricial species include songbirds (suborder Passeriformes), mice (family Muridae), rabbits (genus *Sylvilagus*), cats, and dogs; examples of precocial species are ducks (genus *Anas*), hares (genus *Lepus*), deer (family Cervidae), and cattle (family Bovidae). In mammals, precocial young also tend to (1) be small in body size, (2) be subjected to heavy predation pressure, (3) be born in large litter, (4) be raised in a "nest" with short lactation, and (5) exhibit rapid growth.

A dichotomy between altricial and precocial young is probably simplistic with there being as many as six modes of young development (Hinde 1970; Brown 1973). For instance, the emperor penguin (*Aptenodytes forsteri*) is semiprecocial, with young being mobile yet staying in a nest. There also does not to be an evolutionary trend from being first precocial and then later altricial. In addition, young development is similar in very different phylogenetic groups (hares versus deer) or different in very similar groups (rabbits versus hares). Precocial development of young is confined to an extremely short time period. For example, this critical period or sensitive period may be only 13–16 h of hatching in bird; it is often irreversible and permanent; and in birds, color, pattern, sound (call of mother), and movement seem to be the major stimuli in a young identifying its mother (Gill 1990). In mammals, olfaction (or smell) seems to be the major stimuli (Pough et al. 2002).

Imprinting may be separated into two components (Brown 1973). In the first, filial imprinting, a young animal follows an object (usually a parent). This following keeps the young with the mother after they move away from the den or nest; also, it helps distinguish parents from other adults, which might attack them, and tends to occur in all precocial young. In the second, sexual imprinting, some animals, like the brown-headed cowbird may not see its mate until courtship with a mate. A problem exists in propagation programs, including hacking programs for raptor species of concern. For example, when raising chicks of peregrine falcon (*Falco peregrinus*) or California condor (*Gymnogyps californianus*) in captivity for reintroduction into the wild, models with surrogate heads of a parent are used so that the feeding chick does not confuse humans with food (Dzialak et al. 2006), resulting in possible filial or sexual imprinting on humans.

Compared to studies of imprinting in birds, relatively few studies have been conducted on imprinting-like processes in mammals (Eibel-Eibesfeldt 1970). In mammals, if a young does not recognize its parent, conspecifics of another sex, neighboring conspecifics, or predators could result. In mammals with hooves, e.g., ungulates, young can be imprinted easily on humans, as in the domestic horse (*Equus caballus*). Mammalian females may have a maternal attachment to their young by "labeling" them directly with licking and indirectly with milk (Goodenough et al. 2001). Hence, mammalian young may have to stay with the female for up to 8 h before the female recognizes her young.

4.7 Lactation in Mammals

Mammals are named because of the presence of mammary glands in females (Vaughan et al. 2000). Provision of milk by females, termed lactation, typically requires much more energy than gestation, a term given to the development of young within the uterus of the female.

Lactation is especially costly for a metatherians (or "marsupial"), such as the common opossum (*Didelphis virginiana*), because this opossum has a very short gestation period, with the young born very small and underdeveloped, compared to a common muskrat (which is a eutherian, *Ondatra zibethicus*) (Yahner 2001). Both mammals are similar in size (about 1,300 g); the common opossum has a gestation period of about 12 days, and young are lactated for about 50–60 days; in contrast, the common muskrat has a gestation of 28–30 days, and a lactation period of about 15–16 days (Gardner 1982; Perry 1982). In this comparison, metatherians invest little energy in pregnancy, but considerable energy in lactation; the reverse is true of eutherians.

4.8 Pouches, Parental Care, Locomotion, and Altricial Versus Precocial Young

Perhaps a poorly known fact is that only about 50% of metatherians have pouches, as commonly seen in kangaroos (family Macropodidae) (Vaughan et al. 2000). In fact, being pouchless as a metatherian is considered a primitive condition.

Pouches are best developed in metatherians that hop (kangaroos), climb (phalangers or cucuses, family Phalangeridae), or dig (wombats, family Vombatidae); in the wombat, which are digging, burrowing metatherians, the pouch opens backward!

Kangaroos hop, termed saltation, and are bipedal, rather than running like quadrupedal ungulates, such as comparable sized ungulates. Saltation in kangaroos seems to be related to the energetic costs associated with locomotion, parental care, and perhaps modes of foraging. In the kangaroo, energy is stored in elastic fibrous tissues of the hind legs, which are tendons. Thus, is takes less energy to hop than to run above 9 kph when using bipedal locomotion. However, at low speeds (less than 4 kph), bipedal locomotion is apparently slow, clumsy, and energetically costly to kangaroos. Therefore, perhaps running never evolved in kangaroos or maybe there are fewer predators of kangaroos. These two explanations are unlikely. Perhaps saltation has evolved because kangaroos have a slower metabolic rate than comparable size ungulates (about 70%). A more likely reason for the evolution of saltation in the kangaroos is that young at birth need their forelimbs to climb from uterus to the pouch for lactation. Once a forelimb is adapted to clinging and climbing, it unlikely will evolve into a forelimb for running or locomotion in general.

Kangaroos and large ungulates are ecologically convergent, which in this case means they have evolved in similar habitat and eat similar foods. Large ungulates in open grasslands, however, tend to be highly social, living in rather large groups; in contrast. kangaroos are not highly social and live in smaller groups (usually only 1–2 animals). Large ungulates show coordinated grazing on open plains, but pentapedal locomotion in kangaroos makes it hard to stay together and feed in unison.

4.9 Duration of Parental Care and the Timing of Dispersal

Dispersal is a process, whereby young move from a birth (natal) site to new place. When and how dispersal takes place has been a subject of much controversy and research; the timing of dispersal has relation to the existence and the duration of parental care. In many species of amphibians and reptiles, parental care is rare or nonexistent (Zug et al. 2001). In larger birds, such as terns (family Sternidae), even when young have left the nest, parental care may continue until the art of capturing prey is mastered. In the large swan geese (*Anser cygnoides*), vigilance and aggressive behavior declines in both parents as brood-rearing progresses; instead, juvenile of both sexes increase vigilance and feeding (Randler 2007). In northerly latitudes, the matriarch group breaks up when young of year born; e.g., yearling females disperse; but in southerly latitudes, young stay longer with the mother, perhaps as a function solely of more open landscape (Hirth 1977). In marmots of western states, such as the yellow-bellied marmot (*Marmota flaviventris*), young stay within the colony for an extra year to gain sufficient body size. Eastern chipmunks disperse soon after immergence from the burrow system; these young need to gain body weight for winter (those of 2nd litter) or for breeding (those of 1st litter) (Yahner and Svendsen 1978).

Chapter 5
Dispersal and Corridors

5.1 Timing of Dispersal

When is it time for the young to leave the family unit (Horn 1978)? There is a misconception that parents physically evict or chase away young from the natal site. However, there is very little or no evidence for this. In eastern chipmunks, when young emerge from an underground burrow systems for the first time, play among siblings may initially be seen, but this behavior wanes in a couple of days. Soon the young are intolerant of each other, and the female parent soon becomes very aggressive and territorial toward the young, if the young try to return to the natal site (Yahner 1978a). Thus, if the adult female shows any aggression toward her young, it seems to be simply an intolerance of their presence as if they were an intruding conspecific.

In mountain lion, the female parent simply abandons the young near the boundary of the home range (Beier 1995). In the burrowing owl (*Athene cunicularia*), juveniles in Florida disperse when the burrow is flooded; dispersal distances may range from 330 to over 10,000 m (Mrykalo et al. 2007).

5.2 Reasons to Disperse or Not to Disperse

Recall that dispersion is a distribution (random, clumped, etc.), whereas dispersal is a process (Yahner and Mahan 2002). Dispersal may be defined as the movement by young animals away from their natal site; so it is not, which is seasonal and directional, as with spring migration of songbirds northward. A distinction also can be made between natal dispersal and natal philopatry, whereby an animal stays in its area of birth. So why disperse? Dispersal may reduce inbreeding, which can be deleterious. Perhaps competition with resident adults for food, home sites, or mates

Fig. 5.1 The gray wolf, or timber wolf, is very social, living in packs, which often are closely related. The gray wolf shares a common ancestry with the domestic dog

can mitigate via dispersal. For instance, in white-tailed deer from Texas, young deer disperse as 2.5-year-olds because of aggressive behavior and sexual competition with older males (Hirth 1977). The reverse of dispersal is being philopatric. If philopatric, some inbreeding may be good if it maintains well-adapted genes. Moreover, a philopatric young animal may already be familiar with the location of food and other resources in the area of the natal site (Fig. 5.1).

There seems to be a dichotomy in terms of the predominant sex class that disperse in mammals and birds, with males typically dispersing in mammals and females generally dispersing in birds. Why the possible bias? Mammals are mainly polygynous, so young males better off leaving an area rather than competing with resident males. Birds, on the other hand, are usually monogamous; therefore, a young, male bird may compete for resources and a territory. By staying philopatric, a young, male bird may be to be more familiar with a given area known to them, making philopatry adaptive and dispersal typically lacking in this phylogenetic group.

Animals may disperse passively, as with barnacles (subclass Crustacea) that may piggy-back on the sides of ships, via wind or water currents (as with gypsy moth larvae, *Lymantria dispar*, Fig. 5.2), or directly via actual movements (as with adult gypsy moths). Landscape resistance is a term given to the degree to which dispersal, or movements in general, across a landscape is impeded by barriers, such as highways, developed areas, or gaps in a habitat. I do know of studies that put numbers to landscape resistance; e.g., 6.0 for highways vs. 16.8 for clear-cuts. Several studies, however, have shown that certain barriers can be relative to certain species. For instance, roadways have caused high mortality for aquatic turtles by distorting sex ratios. On the other hand, roads probably do not impede dispersal in mobile carnivores, e.g., bobcat (*Lynx rufus*) (millions and Swanson 2007).

5.2 Reasons to Disperse or Not to Disperse 37

Fig. 5.2 The gypsy moth is a forest pest, from Europe and Asia, having escaped into the USA about 1860. Egg masses of gypsy moths are typically placed on branches and trunks of trees. Egg masses are buff colored when first laid but may become bleached over winter

5.2.1 Corridors in the Landscape

Corridors are connections of isolated habitat patches that may facilitate dispersal. Connectivity is the degree to which a landscape resembles a pristine landscape or the degree to which animals can make movements between isolated patches. The fact that corridors help to better connect a landscape allows scientists to "sell" the concept of a corridor to the public. After all, a relatively homogeneous landscape is what the European settlers encountered when they first colonized North America (Yahner 2000). Second, a corridor may sell to the public because it resembles a paved sidewalk used by pedestrians. In other word, if a corridor (e.g., a "sidewalk") is present, animals are expected to use it. However, there are some unanswered questions regarding the use of corridors by animals. First, do animals use corridors during dispersal? I spent 3 years trying to detect juvenile dispersal in eastern chipmunks to no avail. Not only did dispersal of chipmunks seem to be restricted to only a few days at most, but juvenile animals seem to have a "one-track mind;" in other words, juveniles seemed to be interested only in getting to an unoccupied, yet suitable habitat, and live traps with bait were unimportant to them. I was left with the conclusion that a habitat served as a corridor without any proof of it being used by the animal in question!

A second set of questions regarding a corridor might be do animals use a corridor actively or passively? What constitutes a good or a bad corridor to the species in question? Why is this important to know? Many reasons, among them is cost to acquire connectivity via corridors in the landscape or to create a good corridor (Yahner 2000).

A third set of questions regarding a corridor should perhaps focus on the dimensions and the composition of a corridor. A positive aspect of corridors is that it could allow some forested species to move between two isolated woodlots, provided the corridor also of sufficient width and was wooded; in midwestern states, we found that white-footed mice (*Peromyscus leucopus*) and eastern cottontails (*Sylvilagus floridanus*) had limited dispersal abilities, but needed wooded fencerows amid extensive agriculture (Yahner 2000). Similarly, American robins (*Turdus migratorius*) and brown thrashers (*Toxostoma rufum*) were more likely to move between connected woodlots (Haas 1995). A corridor may be perhaps only be 2 m in width for these four species to facilitate dispersal, but for larger species, the width of a corridor may need to be at least twice wide as the typical diameter of the home range (Lindemayer and Nix 1993).

Among wildlife, birds have been best studied; but even with this amount of study, the optimal width of a corridor for this taxonomic group has varied. For instance, in wooded landscape, birds readily choose 100-m wide corridors over nearby clear-cuts for dispersal (Yahner 1997a). Corridors of over 400 m in width have been recommended for forest-interior birds in Alaska (Kissling and Garton 2008). In southern states, 500-m wide corridors may be needed for woodland birds (Kilgo et al. 1998). Corridors with only a 25-m width have enabled predators of bird nests to move among patches; this connectivity reduced nesting success of an edge bird species, the indigo bunting (*Passerina cyanea*); daily survival rate of nests of this bird were only about 50% in habitats with these corridors compared to habitats without corridors and presumed access by predators (Weldon 2006).

Conceivably, width needed for a given species may vary with foraging strategy, such as an herbivore versus a carnivore (Yahner 2000). Moreover, width of a corridor may vary with the length of corridor. For example, widths of corridors for the mountain lion should be at least 100 m if its length is less than 800 m (Beier 1995). However, width of a corridor designed for the mountain lion should be greater than 400 m, if the length is between 1 and 7 km long.

For open-country species, e.g., butterflies (order Lepidoptera), corridors need not be wooded, rather open for flight (Sutcliffe and Thomas 1996). In northeastern states, where early successional habitat is being lost when the forest matures (Yahner 2004a), as in northern Pennsylvania, electric transmission rights-of-way are extremely valuable to many bird species of concern (Yahner et al. 2003). One positive aspect of corridors is that it could allow some forested species to move between two isolated woodlots, provided the corridor also was wooded and of sufficient width.

Hence, similarity in vegetation between a corridor and the two habitats connected by the corridor is important. The greater the vegetative similarities among corridors and the two connected habitats, the more likely the corridor is likely to be used by wildlife. The greater the similarities, the more likely a given corridor will be used as an alternative habitat, which may be a second value of a corridor (if not solely for dispersal). Woodland birds, such as red-eyed vireos (*Vireo olivaceus*) apparently use 100-m wide wooded corridors as habitat than "isolated" 1-ha (100 m × 100 m) wooded stands (Yahner 2003). Furthermore, a dissimilar matrix in which a corridor and the connected habitats are impeded may not be as important as the quality of a distant habitat. For instance, the quality of a distant pond (230–1,220 m away) was much more important to painted turtles (*Chrysemys picta*) compared to the quality of a closer pond that had actually dried up (Bowne et al. 2006).

Simply looking at the width and character of a corridor often is inadequate for certain species, whose behavior is affected by humans or human activities. For example, in the West, a corridor is of little value to dispersing juvenile (<18-month-old) mountain lions if the corridor contains human dwellings (>1 dwelling/16 ha), has artificial lighting, lacks woody cover, and lacks an underpass at highways that traverse natural travel routes (Beier 1995). Simply the creation of recreational walkways through natural areas can affect the distribution of terrestrial salamanders (Davis 2007).

Waterways, because of their linear nature, must be viewed as potential corridors because of their value to aquatic organisms (Neely and George 2006). The greenside darter (*Etheostoma blennioides*) has become established in the Susquehanna drainage, having been first discovered in this drainage in the 1960s. With 60 native species and 34 exotic species, a question which has risen is whether this species is exotic, having been introduced by anglers via the bait-bucket method. Apparently, this darter has small home ranges and poor dispersal capability. So, whether dispersal is facilitated by the drainage in the greenside darter is difficult to determine. The impact of this darter on other species is unknown, although there is a potential for dietary overlap, and the greenside darter if very tolerant of siltation and nutrient runoff compared to other darters.

Populations may not always be contiguous or connected by corridors. In fact, small populations may go extinct locally and be recolonized by individuals that disperse from neighboring populations. These populations, termed meta-populations, seem to resemble blinking lights on a Christmas tree. When a light goes off (like a population going locally extinct), it later comes back on (as the area becomes established by dispersing animals). Examples of meta-populations may include those of the New England cottontail (*Sylvilagus transitionalis*; Litvaitis 1993) or the Allegheny woodrat (*Neotoma magister*; Balcom and Yahner 1996). For instance, meta-populations of New England cottontail, which is a candidate for federal listing as threatened or endangered, use clear-cuts. When a forest is managed via silvicultural techniques, each clear-cut should to be 15–75 ha and less than 1 km from another clear-cut to better ensure connectivity and dispersal by this species.

5.2.2 Landscape Linkages or Megacorridors

Landscape linkages are often called megacorridors because of their large size. A well-known linkage is that of Pinhook Swamp on the northern border of Florida, which connects Osceola Swamp (to the south in Florida) with Okefenokee Swamp (to the north in Georgia) (Harris 1984). This linkage may play a major role in long-term survivorship of focal species, like black bear (*Ursus americanus*), Florida panther (which is the easternmost population of mountain lion), and red-cockaded woodpecker (*Picoides borealis*).

Another landscape linkage in southern California, where there are 15 major linkages for megacharismatic species, like mountain lion (Beier 2007). Another linkage is being designed in the Allegheny National Forest in northwestern Pennsylvania, which will consist of 33,300 ha of wooded corridors to link eight old-growth stands. Any landscape linkage needs to be more than simply a means of moving animals from one place to another, but it should provide habitat as well. A landscape linkage should avoid urbanization, integrate land acquisition and highway mitigation, and avoid predator control or inappropriate artificial light pollution (Rich and Longcore

2006). Landscape linkages can be important to landbirds, as potential stopover points to feed and rest during migration.

With the exception of spectacular migrations of caribou (*Rangifer tarandus*) in the Arctic, the most important migrations of land mammals are probably those that occur in the Greater Yellowstone Ecosystem (GYE) (Berger 2004). About 75% of the original migratory routes of elk (*Cervus canadensis*), bison (*Bison bison*), and pronghorn (*Antilocapra americana*) in the GYE have been lost. Loss of megacorridors for these large mammals is attributed to little tolerance by landowners outside of protected areas, a concentration of elk on 23 wintering grounds in Wyoming, a 20% increase in human population in the last decade in the region, and a critical loss of habitat. Thus, there is a future challenge of protecting the remaining routes from development, especially public leasing of lands along the routes for energy development.

Chapter 6
Food-Acquisition Systems

6.1 General Comments

All organisms are tied together in a number of ways, with energy acquisition being paramount. Some organisms are meat-eaters, or carnivores, that feed on plant-eaters, or herbivores. Most species, however, are omnivores, feeding on both animals and plants. Of the many foods available to a given animal, which are selected and which are not? Decisions may vary and be related to food availability, ease of capture, or nutritive value. Except when young (Yahner 1978c), an animal cannot spend considerable time sampling food; there are limits in time and energy involved with food capture, handling, and ingesting.

Food for animals often or always is patchily distributed or not be present. This is very different from humans going to a food store to buy predictably a supply of bread or milk. At least in our country, we always know that these foodstuffs will be in the food store. If you ever watch animals, you may find that they spend most of their time foraging for food (Yahner 1980a). If time is diverted from finding food, an animal may have to spend less time mating or detecting predators, so it is a balance (time budgets) that is constantly being achieved (Mahan and Yahner 1990). If some wintering birds of temperate climates do not gather food in the morning, fat reserves depleted overnight can result in starvation (Morse 1970). In fact, many structures (hooked or long beak) of dentition (e.g., compliment and types of teeth) have evolved to maximize food capture and processing.

Filter feeding for food predominates in invertebrates (Goodenough et al. 2001), but this mode of food finding is found in some vertebrates, such as baleen whales (suborder Mysticeti; Vaughan et al. 2000). In all cases, filter feeding is very passive. Baleen whales are the largest mammals and have huge mouths, with baleen (plates, comb-like) suspended from the maxilla (e.g., upper jaw, $n=350$ or so). Baleen whales filter food items (often are less than 1 cm) by slowly swimming through the water or by other means. A bivalve mollusk (class Bivalvia) is also not a major stalker of prey! These aquatic animals passively filter food particles from the water by pumping water across feathery gill structures and mucous membranes.

Herbivores are widespread among animal species, e.g., the gypsy moth, which is a principal exotic forest pest in the eastern USA (Yahner 2000). This moth was brought into Massachusetts in 1869 by a French scientist, who was interested in the silk industry. The gypsy moth subsequently escaped and feeds extensively on oak (*Quercus* spp.).

Leaf cutter ants (family Formicidae) cut fresh leaves and carry the pieces to their underground nests (Goodenough et al. 2001). These ants cultivate a fungus, which is specific to these ant nests, on the leaves, and the fungus serves as a food source for ants. Ants use this fungus garden, by licking the leaves to remove waxy layer on the leaves; ants also chew leaves to a pulp, which is a source of nourishment for the fungus.

Bees (superfamily Apoidea) are among the animals that feed on nectar of plants. There is an increasing concern for the loss of animals that feed on pollen or nectar and no longer are available for pollination (Yahner 51). Honeybee colonies, for instance, are being lost because of inclement weather, parasitic mite syndrome, which affects the trachea, causing stress and acting as disease vectors, and other factors.

White-tailed deer rely on browse in winter and herbaceous plants in spring and summer for food (Scott and Yahner 1989). Deer prefer pin cherry (*Prunus pensylvanica*) and ground vegetation (e.g., raspberry; *Rubus* spp.), allowing less preferred species, such as American beech (*Fagus grandifolia*) and black cherry (*Prunus serotina*), to proliferate; ferns and grasses often dominate forest stands.

6.2 Wildlife and the Prey Rat Race

I recall watching on television a documentary that dealt with the history of the development of the modern tanks in human warfare. In many ways, the tank became a better weapon as warfare became more sophisticated. Similarly, a wildlife and prey rat race is ongoing, with carnivores become more sophisticated in capturing their prey while the prey is becoming more sophisticated in not becoming a meal for a carnivore (Goodenough et al. 2001).

How do carnivores become more sophisticated? A species of baleen whales, the humpback whale (*Megaptera novaeangliae*), traps its food by blowing bubbles, sometimes termed bubble net feeding, at a depth of about 15 m (Earle 1979). The bubbles forms a cylindrical net that traps small food items; then humpback whales swim up and gulp the concentration of prey in the net.

Probably the most familiar trap is the web created by orb-weaving spiders (Penny and Ortuño 2006), and each of us has come across a web produces by one of these spiders at one time or another. Some orb-weaving spiders, such as the oval St. Andrew's cross spider (*Argiope aemula*) builds conspicuous decorations on its webs structures act to lure prey but may increase predation risk to the spiders (Cheng and Tso 2007).

Orb-weaving spiders are examples of sit-and-wait predators (Morse 1970). This type of predator uses ambush, and is usually true of smaller predators. This may partly be true because a sit-and-wait predator requires concealment; if a sit-and-wait predator is large or is not cryptic, prey can easily detect it. A large cat, the tiger

(*Panthera tigris*), may hide in the brush, allowing its stripes to make it difficult to see. Also, this technique is often noticed when prey is much more abundant than that of predators. Sit-and-wait predators typically remain motionless, in one spot, and make quick dashes for prey.

A second general way of capturing prey is by chasing prey (Morse 1970; Ewer 1973). Chasing prey is best shown by members of the dog family (family Canidae). Social dogs, like the gray wolf, chases dangerous prey in a group perhaps to minimize injury to the predator. Cooperative chasing allows these predators to attack large prey; gray wolves can run distances of up to 8 km. These predators tend to specialize on large, which often are young, old, or ill animals. Not all dog feed on large prey; with the exception of the near-extinct red wolf (*Canis rufus*), most wild dog species scavenge on carrion, if available. Smaller dog species, like the red fox (*Vulpes vulpes*) is opportunistic by feeding on a variety of prey items, ranging from small mammals, carrion, and a variety of insects. The largest marsupial (infraclass Marsupialia, family Dasyuridae) carnivore, the Tasmanian devil (*Sarcophilus harrisii*), also may scavenge for food.

In the wildlife–prey rat race, predators may develop sensory specializations for prey detection. To name a couple of these specializations, pit viper snakes, like timber rattlesnake (*Crotalus horridus*) and copperhead (*Agkistrodon contortrix*), have special sensors (pit organs) below the eyes that can detect infrared (heat) in darkness (Zug et al. 1991). These sensors are capable of "feeling" the body heat of an endotherm, such as a mouse, which is within 40 cm. The duck-billed platypus (*Ornithos anatinus*) of eastern Australia and neighboring Tasmania is one of the few egg-laying mammals. This mammal has electrosensors and mechanosensors in it bill for locating prey in murky water at night (Vaughan et al. 2000).

Hunting cryptic prey can be difficult for a predator. In the western USA a reason given for black bear (*Ursus americanus*) with nonblack coloration is that is acts as camouflage from predators of black bears, e.g., brown bears (*Ursus arctos*) or packs of gray wolves (Rogers 1980). If prey is hard to find, a predator may develop a search image for the prey. I found this probably to be true of American crows (*Corvus brachyrhynchos*) in earlier studies of artificial nests (Yahner and Wright 1985). The concept of a search image, which was developed by Tinbergen in 1960, observed in insect prey used by birds in Dutch woodlands (Krebs 1978). Hunting via a search image can be viewed as a constraint or as an adaptation to a predator. It can be a constraint if prey is present for which predator has no search image, thereby the prey is effectively not encountered. However, a search image can be an adaptive if the prey is present; a predator can then maximize its use of time in finding the prey.

6.3 Optimal Foraging Theory

According to optimal foraging theory, it is in the best evolutionary interest of a predator to make the right decisions on which prey type to select, where to forage, and how to forage (Krebs 1978). A prey item might vary with a predator: a prey item

Fig. 6.1 The great horned owl has females that are larger than males. Its "horns" are not ears or horns, but tufts of feathers. It is believed to occur from Nova Scotia to east Texas and to Minnesota and southward to South America. Great horned owls breed early in late January or early February

to a house finch (*Carpodacus mexicanus*) might be a seed, whereas prey to a bumblebee (*Bombus* spp.) might be the nectar of a flower. Efficient predators should be selected in order to maximize energy procurement, which allows time for other things, such as reproduction or detecting predators. Efficiency in predation could mean maximizing the rate of food intake, nutrient intake, or both over a long time period. According to optimal foraging theory, individuals that forage more efficiently will leave more offspring, but this relationship is hard to prove because fitness is a lifetime measure, whereas foraging is usually measured immediately.

Each predator should choose "profitable" prey (Krebs 1978) because each item has a net value (in terms of calories), where gross value is the time it takes a predator to find, capture, ingest, and digest prey. A preferred prey size may dictate its use by a predator. A juvenile American alligator (*Alligator mississippiensis*) feeds on smaller prey, whereas adult alligators feed on larger prey (Zug et al. 2001). Even if a prey is very profitable, it may be too scarce to hunt. Thus, a predator might opt for a more abundant, yet less profitable prey. An example of this trade-off may be shown experimentally by the bluegill (*Lepomis macrochirus*). If the bluegill is given prey (*Daphnia*, order Cladocera) of three sizes, each at a low density (at 20 each), bluegills select prey in proportion to encountering them. However, if the two largest prey are presented to bluegills at high density (at 200/size class), bluegills selected the two largest sizes.

Two predictions stem from optimal foraging theory. The first is that the acceptability of a preferred food item depends on its abundance; it the second-most preferred item is instead the most abundant, a predator will switch to the second-most preferred item (Yahner 2000). This is termed as a functional response, and it occurs, for instance, when snowshoe hare (*Lepus americanus*) become scarce, great horned owl (*Bubo virginianus*) (Fig. 6.1) switch to feeding principally on ruffed grouse (*Bonasa umbellus*) (Rusch et al. 1972). A second is that of a numerical response (Yahner 2000). Gray wolves showed a numerical response by having fewer young when the availability of moose (*Alces alces*) declined, which represented the major prey item of wolves (Peterson and Page 1988).

On some occasions, energy needs to be balance with nutrient content of food. An example is the moose, which needs to obtain appreciable food because of its large size but it also has a minimum daily requirement for sodium (Risenhoover and Peterson 1986). A moose eats both land and aquatic plants because aquatic plants have considerable sodium content. In winter, however, sodium-rich plants are unavailable to lakes because of ice cover; therefore, moose store sodium in their body for use in winter.

Each patch or habitat can have different prey and different abundances of each prey to potential predators. For instance, common raccoons (*Procyon lotor*) are more likely to be found in cornfields in the autumn where corn (prey) is located (Compton 2007). In summer, common raccoons seldom used these cornfields. The importance of a patch or habitat may change seasonally with foraging birds. Wintering birds rely on rough-barked trees as foraging substate. Crevices in the bark of mature oak trees likely contain numerous arthropods as prey (Yahner 1987).

Patches or habitat for one species may not serve for foraging purposes for other species, perhaps because of prey type. Common raccoons forage more often along forest edges adjacent to streams or agricultural fields, whereas no relation with movements of other predators, e.g., striped skunk (*Mephitis mephitis*) and Virginia opossum (*Didelphis virginiana*) (Dijak and Thompson 2000). Logging of a forest stand reduces the amount of foraging substrate for birds (Franzreb 1983). In a national forest of Arizona, mountain chickadees (*Poecile gambeli*) foraged lower in trees and more on aspen (*Populus* spp.). On the other hand, ruby-crowned kinglets (*Regulus calendula*) foraged in smaller trees, but only in trees with dense foliage; other species, e.g., yellow-rumped warbler (*Dendroica coronata*) was more generalized in use of habitat in the logged area. In some cases, whether a species forages in an area may depend on the degree of disturbance on the area. During studies of artificial bird nests, a predator, the American crow preyed on only 9% of the nests in an uncut forested landscape, but preyed on 25% of the nests in a forested area affected by clear-cuts (Yahner and Scott 1988).

Conceivably, food can be found under two conditions, at a site with moderate but constant abundance or at with fluctuations that go far above and below moderate abundance. While foraging, wood storks (*Mycteria americana*) require shallow wetlands with concentrated prey. In wood storks of South Carolinian and Georgian wetlands, birds foraging in tidal freshwater wetlands in Georgia had captured 3 prey items/min compared to only 0.10 prey items/min at nontidal site in South Carolina. But in Georgia, wood storks (Fig. 6.2) were limited to foraging only at low tide (tidal=14 min on average; nontidal=88 min). This suggested that tidal site, e.g., those studied in Georgia, were higher quality foraging areas for wood storks than the nontidal sites in South Carolina. This, study showed that the Georgian tidal sites had abundant, concentrated prey, whereas the South Carolinian nontidal sites had sites with fluctuating prey resources.

Fig. 6.2 The wood stork, or sometimes called the wood ibis, is the only stork that breeds in North America. It extends its range throughout South America, in tropical and subtropical regions

6.4 Central Place Foraging and Hoarding of Food

Some species exhibit central place foraging (CPF) (McFarland 1999), whereby food is taken to a centralized home site, e.g., the underground burrow system of an eastern chipmunk. In contrast, some animals consume prey at the place of capture. Winter in temperate latitudes represents a season of low food supplies for most animals; thus, animals in this season and latitudes have three survival strategies: store food (cache food) for use during winter, migrate to a warmer climate, or become inactive (e.g., hibernate) and live off stored body fat (Yahner 2001). These latter two strategies will be dealt with later.

Hoarding or caching food can be divided into two types, larder and scatter hoarding (Yahner 2001). Larder hoarding is a classic case involving central place foraging, when an animal goes to the foraging area, collects a food item(s), and takes this food item to a home site. Scatter hoarding, in contrast, is the placement of one or a few food items in scattered locations, as with a gray squirrel (*Sciurus carolinensis*) on the forest floor. Larder hoarding, rather than scatter hoarding, generally has evolved in solitary, territorial species (Smith 1968). Territoriality as a spacing mechanism will be discussed later, but it typically is focused around a fixed point (e.g., home site) and involves aggression against an intruder (Brown 1973).

Hoarding, either larder or scatter, tends to occur just prior to the time it is needed or when food is in excess supply (Yahner 2001). Autumnal production of acorns by oak (*Quercus* spp.) coincides with considerable efforts by eastern chipmunks of larder hoard this prey item prior to a period of winter inactivity, termed torpor. Hoarding is well-studied in mammals, but it is also reported in some birds and invertebrates (e.g., ants and bees); it is not reported in fishes, amphibians, or reptiles. Among vertebrates, scatter hoarding is most common in birds, whereas larder is most common in mammals. Blue jays (*Cyanocitta cristata*), for instance, scatter hoarded 54% of the acorns of pin oak (*Quercus palustris*) in one study up to distances averaging 1.1 km from seed trees. Because of this ability to scatter hoard, some believe that blue jays were responsible for the rapid northward spread of oaks in the eastern USA at the end of the Pleistocene after the glaciers receded (Johnson and Adkisson 1986). American crows have been observed scatter hoarding young eastern cottontails (*Sylvilagus floridanus*) (Shew 2006). Gray squirrels, a nonterritorial species, probably scatter hoards single or small numbers of items throughout their home range despite possible theft by other squirrels. In this species, scatter hoarding presumed to be adaptive because caches are abundant, widely scattered, and inconspicuous; presumably, squirrels find these caches by visual cues (landmarks, e.g., logs), memory, and possibly olfactory cues (Vander Wall 1990). Gray squirrels tend to cache acorns of the black oak group [e.g., black (*Quercus velutina*), northern red (*Q. rubra*)] rather than those in the white oak group [e.g., white (*Q. alba*), chestnut oaks (*Q. montana*)] (Steele et al. 2005). Acorns of the black oak group are less perishable and less palatable to squirrels; white oak acorns quickly form a taproot, which speeds up germination and may reduce predation (Fox 1982). Also, gray squirrels may remove the embryo of a white oak acorn to halt its germination and slow its perishability; in addition, acorns of white oaks compared to black oaks have less tannin, which may inhibit digestive enzymes at high concentrations (1–2 vs. 6–9%, respectively) (Ofcarcik and Burns 1971). Red fox also exhibit scatter hoarding, e.g., mice (*Microtus* spp.), perhaps to secure this prey from other predators, insects, and microbes (MacDonald 1976).

Some species may show both larder and scatter hoarding. For example, the eastern chipmunk may scatter hoard (usually summer) about eightfold less than it larder hoards (usually in fall); why this species shows both forms of hoarding is speculative, but scatter hoarding may be a vestigial behavioral pattern (Yahner 1975). Food caches in chipmunks may be impressive, containing as much as 1,000 g of food (Thomas 1974).

American beaver (*Castor canadensis*) also create a food cache via larder hoarding from about 38° North (southern Ohio and Maryland) (Hill 1982). The cache of a beaver colony is usually a raft consisting of two types of low preference or nonfood items in the upper layers [e.g., alder (*Alnus* spp.)] and preferred food items in the lower layers [e.g., sugar maple (*Acer saccharum*)]. As the raft becomes water-logged, it sinks; even if pond freezes over and upper layers of a beaver dam become ice-locked, the lower layers are in water and accessible to the beaver in winter as food.

We often get tired shopping the same store or mall, but how long should a predator stay in a given patch (store)? As an animal stays longer in a patch looking for a food time, the food item could become scarcer and, therefore, harder to obtain (Morse 1980); alternatively, rich-food resources may become depleted, leaving only poor-food resources. Hence, time spent in a patch should depend on the amount of food remaining and the difficulty (energy costs) in getting to the next patch. But when should an animal make a decision to forage in another patch? According to the marginal value theorem (Goodenough et al. 2001), an animal should spend more time in its current patch as travel time between patches increases or as the average quality of the habitat decreases. In Midwestern prairies, a perennial grass called the royal catchfly (*Silene regia*) (Menges 1991); it is becoming locally extinct because of a lack of predators. In this case, the predator is the ruby-throated hummingbird (*Archilochus colubris*), which is less likely to visit patches that contain less than 100 individual plants compared to those with greater than 150 plants; hence, patch size (prey number) affects the likelihood of visiting in a patch or staying in a patch while foraging.

In addition to deciding what to forage upon or how long to stay in a patch, for instance, a predator need to determine an optimal pattern of foraging (Morse 1980). An optimal pattern for foraging is one in which a path is not crossed or crossing is minimized. As with migratory wildebeest (*Connochaetes* spp.), which is a herd-forming herbivore on the African plains, the rate of revisiting a grassland will depend on replenishment of the resources (Vaughan et al. 2000). The best time to forage is likely in the morning for songbirds because of energetic costs of remaining inactive overnight and perhaps insects less able to escape predation when it is colder. Predatory reef fish concentrate foraging activities during the twilight hours, when prey is most vulnerable because of higher light conditions (Goodenough et al. 2001). However, if prey is scarce or environmental conditions are harsh, as in winter for birds in temperate latitudes, then foraging might be adaptive throughout the day (Rollfinke and Yahner 1990).

6.5 Constraints on Optimal Foraging: Predation and Competition

If an animal does not forage optimally, the optimal foraging theory simply should not be discarded (Goodenough et al. 2001). Perhaps other factors, such as predation or competition, affect the way an animal might acquire food items. Competition will be discussed later. A challenge is to eat without being eaten, in the presence of predators and competitors. For instance, juvenile Coho salmon (*Oncorhynchus kisutch*) live the first 2 years of their lives in streams while feeding on small invertebrates (Dunbrack and Dill 1983); juvenile salmon wait for the food items to swim by and then gulp it. If these salmon move to find food items, they become more susceptible to predators. Also, if food items are big or if the juvenile salmon are hungry, they will travel further distances to capture food times. In a sense, a food item drifting downstream is better if eaten by a juvenile Coho salmon rather than allowing a competing conspecific to get it.

In some sedentary species, like eastern chipmunks, optimal foraging may be affected by landscape changes, e.g., clear-cutting (Mahan and Yahner 1999). In clear-cuts, eastern chipmunks spend significantly more time in pausing and less time in foraging and locomoting than in contiguous forest. Perhaps in fragmented forests, chipmunks had to spend more time looking for predators and reduce their conspicuousness to potential predators in more open habitats than in more closed, uncut forests.

If predation is important, the prey needs to be able of recognizing predators (Fox 2003). In some cases, potential prey needed to be "afraid" of something, particularly if kept in semicaptivity (Yahner 1980). Similarly, in the wild, when gray wolves and brown bears returned to the Grand Tetons in Wyoming, moose and elk lost all or most of their predator-avoiding behavior over several decades without these large predators coexisting with them, for instance, allowing wolves to come within 5 m. However, now prey shows increased vigilance and runs away when these predators are in the vicinity; they learn this antipredator strategy within a year. In Alberta, elk populations placed into original range (a process termed translocation) into the Rocky Mountains learned that wolves were predators within 2 years (Friar et al. 2007).

In the Grand Tetons, moose have begun to have young closer to a road (Berger 2007). At this point, brown bears, which preyed on moose calves, are known to avoid roads. This suggests that young moose are produced near roads to minimize predation. The Barbary lion (*Panthera leo*, which is a large subspecies of the African lion famous in movies as the MGM lion that once lived from Morocco to Egypt in the wild as of the 1920s and became famous as the lion that fought gladiators and ate Christians in Roman times) is being restocked into the wilds of Morocco (Weidensaul 2007). But before it is successfully restocked, it needs to learn how to hunt prey and is afraid of noises (e.g., rustling leaves, running water).

6.6 Foraging and Group Life

The African hunting dog (*Lycao pictus*) is the most endangered vertebrate in Africa; once exist in 34 countries, it is now found in only six countries, each with less than 100 animals (Fox 2003). The recent crash in African hunting dogs was previously attributed to predation on the dogs by spotted hyenas (*Crocuta crocuta*). Now, the crash in the population of the African hunting dog is believed to be caused by behavior (cooperative hunting and breeding) and not necessarily because of predation by spotted hyenas.

A group of hunting group of dogs allows the hunting of hunt larger prey and the successful guarding of carcasses from scavenging hyenas. However, humans have also decreased the pack size of African hunting dogs. With an overall decrease in pack size, groups of African hunting dogs have become more sensitive to predation, competition, and sometimes they cannot leave a baby-sitter behind to care for young when others in the pack are hunting prey. A smaller pack size also affects the way dogs hunt. Different members of the hunting pack of African hunting dogs play different roles in the hunt, with some flushing the prey, some making the initial attack, while others are adept at disemboweling the prey. Thus, a successful hunt requires both numbers and cooperation in African hunting dogs.

Outside of marine reserves (e.g., off coast of southeastern Italy at Torre Guaceto), overfishing of predatory fish (e.g., *Diplodus sargus*) has resulted in increase of adult sea urchins (e.g., *Paracentrotus lividus*); results in rock reefs being overgrazed by urchins, with reefs going from macroalgal beds to barrens (Guidetti 2007).

Both food habit and predator–prey studies are very common in the literature. Foraging strategies or feeding strategies are the subject of many studies that may deal with behavioral, physiological, or morphological adaptations of an animal to handling, consuming, and metabolizing food and nutrients in the environment

(Servello et al. 2005). I believe that considerably more research is needed on foraging strategies; in the past, wildlife biologists assumed that an adequate food supply in the habitat is all that is necessary for a population regardless the distribution of the food and how an animal finds this food. For example, flocks of migratory songbirds in edge-dominated habitats (e.g., forest edge) tended to move slower than in interiors of forest compared to forest edge, e.g., forest edges at 4.7 m/min vs. forest interiors at 5.6 m/min (WB 1994:704). This suggests that food abundance (e.g., arthropods) is higher along forest edges than in forest interiors. Fall migratory birds may be faced with finding adequate food to get to wintering grounds, whereas spring migrants need to get enough food to get to breeding grounds. Body mass gain is greater in fall than in spring, suggesting that food may be less available in spring than in fall; stopover periods tended to be longer in fall than in spring (average of seven species = 3.9 days in fall vs. 2.9 days in spring).

An impact of gypsy moth defoliation is a reduction in mast (acorn) production, which is an important food for black bears (Yahner 2000); in Shenandoah National Park, reduction in acorns due to gypsy moth caused short-term shifts in bear food habits in fall from acorns to grapes, pokeweed, and other soft fruits (Kasbohm et al. 1995). This shift in food habits of black bears had no discernible effect on survival or reproduction. In Vermont, black bears gained weight prior to winter lethargy (see winter strategies in Chap. 16) by climbing only larger, healthier beech (*Fagus grandifolia*) trees to obtain beech mast in an area affected by beech bark disease (trees affected by the insect, the beech scale, *Cryptococcus fagisuga*). But black bears did not shift food resources, yet they changed foraging habits.

Knowledge of foraging strategies is important to developing sound habitat management strategies; an example may be the development of habitat management for mountain sheep (*Ovis canadensis*) in Colorado (Risenhower and Bailey 1985). Sheep seem to prefer open habitat, spending 11% of their time in this habitat because of a high density of forage; mountain sheep are herbivores. Grassy openings, however, comprise only of 2% of the available habitat. Grasslands provide food, but they also minimize obstructions in seeing approaching predators. Ewes (adult females) and rams (adult males) were more alert (12–15% of time) than were juveniles (less than 6% of time) while foraging. Ewes with lambs also were even more alert. Sheep tended to forage in groups of ten or more animals, perhaps to increase predator detection. Also, foraging efficiency declined with greater distance from escape cover. Thus, optimal habitats for mountain sheep seem to be large open areas near escape cover (higher terrain).

6.7 Predation and Prey Distribution

Besides competition, which will be covered later, predation is a principal biotic factor affecting distribution and perhaps abundance of prey. Browsing by white-tailed deer can affect forest regeneration (Yahner 2000). Because of this browsing, browse lines on woody vegetation may be evident from ground level to about 2 m above ground; 2 m represents the height of foraging deer. Deer browsing partially may be affecting regeneration of tree, e.g., northern red oak (*Quercus rubra*). Forest composition may be altered by deer browsing, thereby allowing less palatable species, e.g., fern (Pteridophyta) to dominate. At Cades Cove, Great Smoky Mountains National Park, deer have changed forest composition from deciduous trees species to conifers (Bratton 1980). In the 1980s, high deer populations at Gettysburg National Military Park, deer grazing and browsing eliminated crop production and forest regeneration (Vecellio et al. 1994). Damage by deer on crops in the USA is estimated at $100 million per year (Conover 1997).

Deer browsing and grazing may affect oak regeneration but may have no effect on beech regeneration (Yahner 2000). Grazing by deer also can negatively affect the abundance of a native endangered species, e.g., the hairy puccoon (*Lithospermum caroliniense*) (Campbell 1993). Near Washington, DC, deer affected the exotic oriental bittersweet (*Celastrus orbiculatus*) but not the exotic garlic mustard (*Alliaria petiolata*). The introduction of elk, which were nearly extirpated in California had reduced the abundance of an invasive exotic grass (*Holcus lanatus*) and decreased native shrubs (Johnson and Cushman 2006).

Deer, elk, or other ungulates are not the only predators affecting distribution or abundance of plants, but insects also can cause this effect. For example, tortrix moths (family Tortricidae) are known to feed on seeds of the northern blazing star (*Liatris borealis*), which is a native and typically a state endangered species (Vickery 2002).

6.8 Humans as Prey

The fact that there is no documented case of gray wolves attacking humans in North America has been termed the "harmless wolf" myth (Geist 2008). A human was killed by wolves in 2005; the kill was thought initially to be caused by a black bear. Rabid wolves historically occurred in Eurasia as early at 2,500 years ago (Yahner 2001). In the early thirteenth century, aggressive attacks by gray wolves on humans were prevalent, perhaps giving us the "big bad wolf syndrome."

In North America, rabies has been reported (Chapman 1978). In areas inhabited by both humans and gray wolves in the USA, wolves will continue to confront humans walking dogs; wolves may even test humans by nipping at clothing, particularly as natural prey becomes unavailable, and potential lethal attacks may follow (Geist 2008).

Compared to gray wolves, there has been a rash of unprovoked attacks on humans from 1890 to 2005, with 19 human fatalities out of 117 attacks (Beier 1991; Sweanor et al. 2008). Most of these attacks have been in western states and provinces that have changed the status of mountain lions from bountied predators to a game of fully protected species. The likelihood of contact between a human and a mountain lion seems to be greatest where humans have encroached on the habitat of mountain lions and humans have increased recreational activity (hiking and biking) in the habitat of mountain lions. In Vancouver, British Columbia, many attacks have occurred by mountain lions on humans because smaller prey is less abundant on the island or because hunting with dogs has resulted in mountain lions becoming very aggressive over time. Most attacks on humans by mountain lions are crepuscular; most humans are on these trails most active during the day. Most of these trails are used by mountain lions; if caches of food are made by lions, they are often placed along these trails.

Probably some mountain lions habituate to humans, resulting in individual differences in cat response to humans (Sweanor et al. 2008). If a human is attacked by a mountain lion, he/she should not be passive in resisting the attack. Because mountain lion attacks usually involve children alone or accompanied by other children or humans with dogs, humans should maintain eye contact, shout, clap hands, and fight back aggressively in all ways possible. It is important not to run away but retreat at a moderate or a slow speed.

Before discussing bear attacks on humans, why are brown bears, i.e., grizzly bears, so much more aggressive toward human than black bears? (Yahner 2001; Herrero et al. 2005). Both bears evolved from the small forest Etruscan bear (*Ursus estuscus*) during the Pleistocene, which crossed the Bering Strait Bridge from Asia into North America. The black bear remained a forest specialist; therefore, the antipredator strategy of a black bear female with young was (is) to climb a tree rather than stand and fight. In contrast, the brown bear became (is) adapted to open, treeless areas left by glaciers, foregoing the need to climb trees; the antipredator strategy of a brown bear female with young was (is) to stand and fight the predator, because after about subadult size, long front claws, powerful shoulder muscles (hence hump), and large body size of the brown bear prevents the ability to climb trees.

Increased aggressiveness in brown bears also may be due to reproductive output, with the brown producing only 6–8 young in a lifetime and the young remain with her prior to dispersal for 2.5 years (Yahner 2001; Herrero et al. 2005). The black bear, in contrast, may produce 12–13 young in a lifetime and young disperse by 1.5 years. Thus, black bear young are more dispensable with less "investment" by the adult female per young over lifetime.

Of brown bear attacks on humans, 70% are related to female defending young, whereas only 30% are food related (Yahner 2001; Herrero et al. 2005). Today, the number of human–black bear interactions is more common than human–brown bear interactions. This, in part, is attributed to greater abundance of black bear versus brown bear in North America. In Yellowstone National Park from 1930 to 1978, there were 2,002 bear attacks on humans, yet only 75 (<4%) were those of brown bear; but of these brown bear attacks, two were fatal (2.6%) compared to one (0.05%) fatality owing to black bear attacks. During this same period, six (20%) of 30 brown bear attacks in Glacier National Park were fatal; in general about 50% of brown bear attacks require at least 24 h of hospitalization. Thus, because of the power of a brown bear, if an attack is imminent, the best strategy to prevent serious injury via passive resistance by playing dead and covering your head and neck area. When humans disturb brown bears, as in Glacier National Park when hiking, brown bears may spend 23% more time acting aggressively toward conspecifics, 52% more time moving, and 53% less time foraging for food. Hence, the mere presence of humans can agitate wildlife.

A disturbing trend in the last couple of decades is that humans now become potential prey for black bears (Herrero 1989). Adult male black bears may methodically stalk humans. This may be recent phenomenon because of human encroachments on bear habitat via human residences or recreation. If stalked and subsequently attacked by a black bear, passive resistance is not advisable. The black bear is smaller than most brown bears so an attack by black bears should be met with all means of aggression, using rocks, sticks, yelling, and clapping hands. A person approached by a black bear should use a slow retreat, climb the nearest tree, and do not run.

Red-pepper sprays are recommended for self-defense against a predator. These sprays contain oleoresin capsicum, which is an irritant. Red-pepper sprays should be dispersed downwind as a cloud rather than as narrow stream. Spraying food with red-pepper spray actually may attract bears because oleoresin capsicum represents a novel smell, which may evoke scent-marking (scent-marking will be discussed later in Chap. 13).

Red-pepper spray stopped an attack on humans by 92% (61 cases) of brown bears (Smith et al. 2008). Of the people using red-pepper spray properly (i.e., not spraying into the wind) against attacks by brown bears, 98% of the people were not injured, and none needed hospitalization. Therefore, proper use of red-pepper spray against attacks of brown bears on humans is an effective method of defense.

6.9 Conservation and Warfare

Since 300 BC, writings have dealt with the impact of warfare on wildlife [e.g., impact on Asian elephants (*Elephas maximus*)] (Dudley et al. 2002). Since 2002, there have been at least 160 wars worldwide. For example, ongoing in Africa (Democratic Republic of Congo) during the 1990s, poaching of great apes [bonobo or pigmy chimpanzee (*Pan paniscus*)] has increased. The Vietnam War in Asia routinely resulted in killing of Asian elephants via bombing and defoliant herbicides, because these elephants were thought to transport military supplies.

Chapter 7
Additional Adaptations Against Predation

7.1 Some Less-Direct Adaptations

As butterflies, monarchs (*Danaus plexippus*) (Fig. 7.1) do not fight predators or act passively if an attack by a predator is imminent; instead, the monarch uses warning coloration and chemical defenses against predation. As adults, monarchs are brightly colored in orange, black, and white; they, therefore, stand out in the environment to a predator, much like the pattern of a poisonous coral snake (*Micrurus fulvius*) (Zug et al. 2001). What does the coloration of a monarch signal to a potential predator? That is unpalatable, so stay away. Monarch becomes unpalatable because of toxins that larvae feed upon in milkweed (*Asclepias syriaca*). These toxins are cardio glycosides, which elicits vomiting in birds that may feed upon them. Only two bird species, the black-headed oriole (*Icterus abeillei*) and the black-headed grosbeak (*Pheucticus melanocephalus*) are known to feed on monarch with no ill effect. In fact, the viceroy butterfly (*Limenitis archippus*) mimics the coloration of the monarch to offset predation on it, which is a form of mimicry known as Batesian.

Some marsh birds, such as the American bittern (*Botaurus lentiginosus*), have simple markings, and they point their beak upward in the presence of a predator (Morse 1980). The theory behind this is that stripes disrupt the outline of the prey as it stands motionless amid the vegetation, thereby confusing the predator. But what about the stripes of a zebra (*Equus* spp.)? On the African plains, vegetation is not present to hide zebras from predators, e.g., African lions (Yahner 2001). Instead, the coloration (or stripes) of zebras acts as a kaleidoscope when zebra are running from predators (Kingdon 1984). Moreover, the black-and-white pattern of zebras stimulates nerve cells in the visual system of prey to help maintain cohesion in a fleeing herd.

Some prey species exhibit countershading, which typically means that animals have darker coloration on dorsal (upper) surfaces and lighter coloration on ventral (lower) surfaces (Goodenough et al. 2001). Hence, when light comes from above, the ventral surface in typically in a shadow, so this portion of the body usually is paler on ventral than on dorsal surfaces, which make animals uniform in coloration. Evidence for

Fig. 7.1 The monarch is perhaps the best known of the butterflies in North America; it is often called the milkweed butterfly. Besides occurring in North America, monarchs are found in Europe and New Zealand. It is famous for its migration in the Americas, especially to Mexico, which may span 3–4 generations

countershading in animals is lacking; a discrepancy in coloration from dorsal to ventral surfaces may be due to either thermoregulation or protection against UV radiation.

Jellyfish (phylum Cnidaria) are transparent with high water content in tissues and few light-absorbing molecules or pigments (Goodenough et al. 2001). In addition, jellyfish typically are small in body size, thereby providing additional crypsis against predators. Why is transparency more common in aquatic than in terrestrial prey? First, refractive indices of light (angle at which light blends when passing from one medium to another) is less in the water medium and in an animal in water, than in an air medium or in an animal outside the water. Second, there may be possible deleterious effects of ultraviolet radiation on land, whereas this radiation is filtered out within a few meters from surface of the water.

Some animals may not be transparent or virtually colorless as antipredator strategies (Goodenough et al. 2001). Instead, some animals may use instant body color change, perhaps as a means of avoiding predation. Well-known example of this color change occur in some lizards, known as chameleons (family Chamaeleonidae); changes in body coloration also are well-developed in cephalopod (family Cephalopoda), which include squid, octopuses, and cuttlefish. One species of cuttlefish (*Sepia officinalis*) is well-known for its ability to change its body color. This cuttlefish also provides the cuttlebone use in pet parakeet or budgerigar (*Melopsittacus undulatus*) to keep the beaks of these birds sharp. When it is swimming, the cuttlefish is brown-and-white for presumed camouflage. But when the cuttlefish is resting at the bottom of the water, it has a coloration that is darker on the bottom, and the color matches that of the bottom.

Unlike cephalopods, which can change coloration very quickly, fox squirrels are considered the most polymorphic mammal. In the southeastern USA, color of fox squirrels in a population or the same litter may vary, with different percentages of gray and black on dorsal surfaces (Fig. 7.2). A hypothesis given for this gray-and-black coloration is that it resembles charred logs on forest floor; this species spends considerable time foraging on the ground, and natural wildfires in southeastern USA are common. Because animals, such as fox squirrels, lack the ability to change color or pattern to match the environment, selection (thus, choice via behavioral means) of the wrong background or improper orientation on the background may become important.

7.1 Some Less-Direct Adaptations

Fig. 7.2 The fox squirrel is closely related to gray squirrels, but fox squirrels are considered the most polymorphic mammal. In the southeastern USA, for example, the color of fox squirrels has different percentages of gray and black on dorsal surfaces

The famous peppered moth (*Biston betularia*) is given usually as the best example of polymorphism. There are two morphs of peppered moths: a white form, with a sprinkling of black dots, and a melanistic form, which is almost completely black (Partridge 1978; Goodenough et al. 2001). The black morph presumably is cryptic against predators on soot-covered trees. Before the industrial revolution in England (in the 1850s), less than 1% of the moths in population were of the melanistic morph; by 1895, however, 98% were melanistic in England. As a result of legislation to reduce industrial pollution and thereby enhance air quality, the frequencies of melanistic forms were reduced in subsequent decades. Presumably melanistic forms were better protected in heavily polluted areas with soot-covered trees compared to unpolluted areas with lichen-covered trees. But in recent years, frequencies of melanistic form have increased in England, suggesting that perhaps predators or parasites of peppered moths are affected by pollution levels.

Animals do not, or at least do not always, select a background; instead, possibly they simply align their bodies in a certain way; e.g., moths on bark with strips of bark, whereby orientation is not with bark coloration but rather with gravity (Sargent 1969). Some predators can even find motionless prey based on postures used by prey! (Nelson and Jackson 2006). For example, jumping spiders (*Evarcha culicivora*) from east Africa feed indirectly on the blood of vertebrates by preying on blood-carrying female mosquitoes as preferred prey, including female mosquitoes (*Anopheles*) that are vectors of human malaria (*Plasmodium falciparum*). The jumping spider selected dead *Anopheles* mosquitoes placed in life-like postures.

Many young animals have coloration to protect them from predation. Freezing and spots of fawns of white-tailed deer probably make the fawns less conspicuous to predators (Mech 1984). Young northern water snakes (*Natrix sipedon*) remain motionless unless movement is necessary, and they also have color banding that provides crypsis (Pough 1976). If young northern water snakes need to move, the movement is very quick. So, for many prey, it is adaptive to remain immobile.

Some species, e.g., black bears in open habitats in western states, are nonblack. In fact, about 50% of the black bears are lighter colored (Rogers 1980). One hypothesis for this nonblack coloration in black bears of western states is that lighter colors in black bears enable them to forage in open meadows at midday. In contrast, black-colored black bears avoid these western open habitats in midday to minimize heat stress. Therefore, a point to be made is that although predation reduction seems to be a primary explanation for polymorphism in a population, there may be another explanation, which may be physiologically based (Goodenough et al. 2001). An ectotherm, the banded snail (*Cepaea nemoralis*), also may have different-colored morphs as a means of increasing heat resistance, which is unrelated to predation risks. As we shall see later, coloration also is important in communication of many animals.

Color patterns in some animals may be a compromise between being cryptic, which would minimize predation, or being conspicuous, which could maximize mating opportunities. In male collared lizards (*Crotaphytus collaris*), for instance, attack frequencies were highest in male with body coloration that best contrasted with the environment. In contrast, inconspicuous females were never attacked (Husak et al. 2006). Thus, there seems to be a rat-race with some animal species evolving conspicuous features to attract mates (sexual selection), yet remain inconspicuous to hunting predators. One species of guppy (*Poecilia reticulata*) seem to have circumvented this dilemma (Endler 1978). In this fish, females select bright males, but predation is reduced in males with fewer and smaller spots and a reduced diversity of colors and patterns in the population. But when predation is low, size and number of spots, and colors and patterns of males become more conspicuous.

Crypsis may not be foolproof, especially when predators develop a search image for cryptic prey (Krebs 1978). Abundant, but cryptic, prey may get around development of search images by occurring in different colors or shapes, or by becoming polymorphic, as with fox squirrels or in peppered moths mentioned earlier. In addition, some species are polymorphic not to escape predation via crypsis, but rather can occur at dramatically different densities as morphs; an example of this "flooding" the predator with morphs of brittle stars (Ophiurida) (Moment 1962). In a sense, morphs appear as different species, with the rarest morph at a selective advantage because a predator learns search image for the more common morph; this is termed apostatic or reflexive selection (Harvey and Greenwood 1978). It apparently is easier for a predator to select the most common morph. Therefore, being a species that is different via morphs may pay off by reducing predation pressure.

7.2 Warning Coloration

Tropical frogs (genera *Phyllobates* and *Dendrobates*) are very brightly colored to warn predators, which is termed aposematism (Goodenough et al. 2001). Native peoples of Columbia (Choco Indians) use these toxins by wiping the tips of their blowgun darts across the back of one of these frogs; some have enough toxic skin secretions to kill several humans or 20,000 house mice (*Mus musculus*).

Aposematism is not limited to tropical frogs; it is seen in bold markings of black-and-white in skunks; bands of black, yellow, and red of coral snakes in the USA; and yellow-and-black in social wasps (Yahner 2001). In addition to bold markings, many of these animals have additional warning signals, e.g., pungent smell of skunks, musk of the stinkpot turtle (musk turtle in eastern USA), or the buzz of a wasp.

The stinkpot family (Kinosternidae) of musk and mud turtles are aquatic turtles, which if disturbed, emit musky secretions from openings on each side of the body (Shaffer 1995). The common musk turtle (*Sternotherus odoratus*) has bold pairs of white or yellow stripes on sides of its head.

The striped skunk makes noises when approached by stamping its front feet, hissing, growling, or clicking its teeth; occasionally, it will also make itself appear bigger by arching its back, standing on its front feet, or walking a short distance on its front feet (Yahner 2001). Use of musk by skunks is the last resort; just before spraying it forms a U-shape with head and rump held high; in this attack position, it pulls its tail back and aims the nipples of the musk glands (modified skin glands) near the anus in the direction of the attacker. A tiny raindrop spray can be discharged in a stream for 2–3 m and up to 5 m. A skunk can spray more than once; the spray causes nausea and burning of eyes and nostrils.

7.3 Mimicry

Mimicry is probably much more widespread in animals than reported, in part, because we observe mimicry in human terms. In other words, if we were capable of detecting other forms of communication, besides visual cues, we probably could describe a variety of mimicry forms. One form of mimicry is termed collective mimicry (Morse 1980). Schooling in some fish is a form of collective mimicry. A school of individual fish may appear as a large prey item, which would deter some predators. However, there is no conclusive evidence for this type of mimicry.

A second form of mimicry, which we view as visual, is termed Müllerian mimicry; it occurs when two or more harmful, but unrelated species are available to a predator (Goodenough et al. 2001). This form of mimicry is named after the German naturalist, Fritz Müller, who proposed the concept in 1878. If a predator tastes one of the toxic species, it will avoid the other; this form of mimicry is typically linked with conspicuous "warning coloration," termed aposematic coloration; presumably Müllerian mimicry is found in tropical butterflies of the subfamily Heliconiinae (Pasteur 1982).

Another, more common, form of mimicry is called Batesian mimicry, which is named after the Amazonian explorer, Henry W. Bates (Goodenough et al. 2001). Probably the best example of Batesian mimicry is that of the palatable viceroy and the unpalatable monarch butterfly, which was mentioned earlier; but unpalatability in monarchs can vary with respect to the amount of alkaloids in milkweed. A palatable species resembles an unpalatable species to predators, e.g., blue jays; butterflies, e.g., the black swallowtail (*Papilio polyxenes*) and the red-spotted purple (*Limenitis arthemis*), resemble the unpalatable pipevine swallowtail (*Battus philenor*). Black swallowtails even roost with their wings closed because ventral surfaces of wings closely resemble those of the wings of pipevine swallowtails; however, in Texas, lizards [anoles (family Iguanidae)] still eat pipevine swallowtail. Thus, like any form of mimicry, it may not guarantee no predation.

An interesting form of mimicry is related to aggressive behavior. As mentioned earlier, about 50% of the black bears in west of the Mississippi River are nonblack (Rogers 1980). Perhaps the nonblack hue or western black bears better resemble the "color hue" of brown bears; by having this hue, a black bear in the distance may be mistaken for another brown bear and be avoided by a wolf pack or by a brown bear (Yahner 2001).

Harmless insects, e.g., some flies, mimic the yellow-and-black bands of wasps or produce the buzzing sound of bees and wasps. An example of this aggressive mimicry extends to an arctiid species of moth (family Sesiidae) (Edwards et al. 1999). Because ants sting or bite and have an unpleasant taste (because of formic acid), they are avoided by many insectivorous predators; therefore, ants have many mimics (Goodenough et al. 2001); in fact, some tropical spiders (family Zodariidae) closely resemble ants with an elongated shape and slender legs (sometime termed myrmecomorphy) (Pasteur 1982). In these cases, a spider mimic may use its front pair of legs to simulate ant antennae that are moved continuously (spiders have four pair of legs compared to three pair in ants).

Aggressive mimicry also is shown probably in clan behavior of spotted hyenas (Vaughan et al. 2000). This hyena sometimes is found often in clans of up to 80 animals, with females being and dominant to males. When two spotted hyenas meet, they go through a meeting ritual involving mutual examination of external genitalia; the external genitalia of a female spotted hyena resemble that of a male. Mimicry in genitalia is believed to allow a "cooling-off" period to mitigate fighting among clan members.

Alligator snapping turtles (*Macrochelys temminckii*) have a wormlike growth on the end of their tongue that they wiggle to attract fish (food item) as they lie on the bottom of a lake with its mouth open (Zug et al. 2001). Phony cleaners (*Aspidontus taeniatus*) are blennid fishes that mimic beneficial blennid fishes that are cleaners [e.g., cleaning wrasse (*Labroides dimidiatus*)]; cleaners remove parasites, diseased tissue, fungi, and bacteria from other fishes (Goodenough et al. 2001). Cleaner fishes "advertise" their presence with distinctive swimming motions above cleaning stations; phony cleaners mimic these motions, then they takes a bite out of fishes going to these stations.

Vocal mimicry is well-known in the northern mockingbird (*Mimus polyglottos*), which is capable of broadcasting songs of many other bird species (Gill 1990). What is the function of this vocal mimicry? Perhaps it helps to exclude other species because mockingbirds are known to chase other bird species, thereby song in mockingbirds may reduce aggressive encounters (functions of bird song will be discussed later). Vocal mimicry may not always serve to reduce aggression, as in the northern mockingbird, but is may help attract other species congregate to help scold and discourage a predator (Dawkins and Krebs 1978; Goodenough et al. 2001). This may be the role of vocal mimicry in a tropical tanager, known as the thick-billed euphonia (*Euphonia laniirostris*) when its nest is threatened by a predator.

Brood parasitism and parental care were discussed in Sect. 4.4. Mimicry in brood parasites also may occur (Gill 1990). Eggs of several Eurasian species of cuckoos often are the same color as those of the host, e.g., blue eggs of common cuckoo match those of a primary host, the common redstart (*Phoenicurus phoenicurus*), with the only difference being that cuckoo eggs are thicker shelled than those of the redstart. Widow birds (family Viduinae) are brood parasites on grass finches (family Estrildidae); young of brood parasites closely resemble those of hosts in terms of plumage and gape markings in the mouth (Eibel-Eibesfeldt 1970).

Butterflies are well-known as having "eye spots" that possibly mimic the eyes of predators (Eibel-Eibesfeldt 1970). Birds have been observed to fear "eye spots" on butterfly wings that are exposed suddenly when in danger (Brown 1970), with spots perhaps resembling the eyes of a snake. Eye spots may startle the predator, particularly if spots are large, few in number, and brightly colored. The black tip on tails of two weasels, the ermine (*Mustela erminea*) and the long-tailed weasel (*Mustela frenata*), is explained in this way; in contrast, the very small least weasel (*Mustela nivalis*) has a very short tail and no black tip. In a sense, the black tip on a tail acts to distract a predator away from the vital head area.

Some animals have evolved false heads to divert predator attacks away from the heads of prey, which may be a form of mimicry (Goodenough et al. 2001). The logic behind this is similar to that of the reason why eye spots are on the outer wings of butterflies or why some weasels have dark spots on their tails. A type of lycaenid butterfly (*Thecla togarna*) has wing patterns, dummy antennae, and dummy eyes at the posterior end of its body to resemble a head, which actually moves.

Autotomy or loss of a "disposable part" of the body (usually a tail) to a predator is reported. This phenomenon is reported as being common in lizards, some salamanders, a few snakes, a few rodents (ground squirrel, *Spermophilus lateralis*), and even in some invertebrates (Zug et al. 2001). Tail loss via autotomy in lizards allows them to get away from the attacker, and it may distract the predator (detached tail may thrash for up to 5 min in some species). Lizards and snakes often have distinctive fracture points in all but the 4–9 anterior-most caudal vertebrae (Pough et al. 2002:318). The tail or lost body part is regenerated, but what is the cost of autotomy? In lizards, the loss of a tail may reduce speed, endurance, swimming, or climbing ability. Juvenile lizards that lose tails have a lower growth rate than those that do not. As with the American beaver (Yahner 2001), the tail also stores fat reserves [60% of fat stored in tails of female geckos (*Coleonyx brevis*)]. Without these energy reserves, reproductive output in some lizards; some lizards may eat autotomized tail because of fat reserves, as in American skink (*Scincella lateralis*).

Sea cucumbers (phylum Echinodermata) forcefully expel their visceral organs (guts) out their cloaca when in danger (Goodenough et al. 2001). A predator may be diverted by these expelled guts and feeds on them while the sea cucumbers make a slow and successful escape.

7.4 Playing Possum and Enhancement

Playing possum is one of the several behavioral concepts that has found its way into popular language; we say that "someone is playing possum" when they are not telling the truth or in some way being deceitful. Yet, feigning injury, or even death, is a widespread antipredator strategy (Goodenough et al. 2001). Perhaps the best-known example of playing possum is shown in the Virginia opossum. The Virginia opossum does not divert the attention of a predator, but it instead "plays dead." Some predators will kill prey only when it is moving, so the opossum playing dead may be ignored by becoming nearly catatonic (no movement, expression) as a defense of last resort.

Playing dead is not restricted to the Virginia opossum. Juvenile caimans (*Caiman crocodilus*) react aggressively toward humans when approached on land, but they feign death when handled in water (Goodenough et al. 2001). Hognose snakes (*Heterodon platirhinos*) have a complex repertoire of antipredator strategies, with feigning as one option; when first discovered, they bluff by flattening and expanding the first one-third of the body to form a hood, which make them appear larger. If this fails, they then curl into an S coil and hiss, while making false strikes at the attacker; if provoked further, they writh violently, defecate, roll over with belly up, and the mouth is opened feigning death (Shaffer 1995; Goodenough et al. 2001).

Ground-nesting birds, e.g., ruffed grouse, when flushed from nests or broods may feign injury rather than death, using a broken-wing display and vocalizations, to divert attention of approaching predator from nest or young (Brown 1970). A parent bird may drag a "broken" wing and flutter away, drawing a predator away from the nest or a brood; then when it sufficiently away, the parent suddenly "recovers;" often then using a "rodent-run display" that involves running in a low crouch, which probably is appealing to mouse-hunting predators, such as red fox.

Some animals, e.g., domestic cats (*Felis catus*), hunch their back and erect their fur when confronted by domesticated dogs. Similarly, a skunk appears bigger before ejecting a spray (Yahner 2001). Some toads and fishes inflate body size (Zug et al. 2001). Domestic dogs bare teeth, and porcupines erect spines to intimidate a predator (Yahner 2001).

7.5 Weaponry in Animals

Stellar's jay (*Cyanocitta stelleri*) is known to break off a twig and use this as a weapon against American crow from a feeding platform by thrusting the pointed stick at the crow. Many deer use hooves or canines, e.g., Chinese barking deer (*Muntiacus reevesi*; Barrette 1977) against predators.

Some ungulates show weapon automimicry, whereby ears are large and near horns; ears often marked or adorned with hair to mimic horns, as in the large, drooping ear tips of roan antelope (*Hippotragus equines*) (Vaughan et al. 2000). Other ungulates may show weapon automimicry by having facial markings that extend the line of the horns onto the face, e.g., sable antelope (*Hippotragus niger*) and oryx (*Oryx beisa*).

Some insects are masters at chemical warfare against potential predators (Goodenough et al. 2001). For instance, assassin bugs (*Platymeris rhadamanthus*) can spit saliva toward an attacker; the saliva is rich in enzymes that cause local pain when it comes in contact with eyes or nose of a predator. Bombardier beetle emit a hot,

7.5 Weaponry in Animals

irritating spray; heat of this spray is produced by mixing chemical reactants mixed in glands (at tip of abdomen) just before it is spraying. Four species of horned toad or lizard (*Phrynosoma* spp.) can eject a stream of blood from its eyes as an antipredator strategy. Although not yet determined, the blood may contain noxious chemicals.

Startle mechanisms, sometimes termed deimatic or distraction displays (Brown 1970; Edmunds 1974), which may be lumped with the broken-wing display of a bird act to surprise a potential attacker. Is it surprising that this is effective against an attacker? Did you ever flush a rabbit or grouse under your feed as you walk in the woods? It takes you, and presumably an attacker, about 1–2 s to recover, which gives the prey ample time to escape. In fact, any good self-defense course, in my opinion, teaches yelling as the first line of defense against an attacker. Tail flagging in solitary animals, like eastern cottontails, probably fit into the category of startle mechanisms (Yahner 2001). Tail flagging will be discussed in a section dealing with visual communication (see Chap. 12).

Some prey species give auditory signals to predators to presumably let the predator know of their location (Sherman 1977; Yahner 1980b). In the wild, vocalizations by Beldings ground squirrel (*Spermophilus beldingi*) deter coyotes from hunting in the area (Sherman 1977); similarly, tigers (*Panthera tigris*) seem to stop hunting Chinese barking deer when these deer vocalize (Schaller 1967).

Stotting by mule deer (*Odocoileus hemionus*) and some other ungulates is a stiff-legged jumping display in which all feet are off the ground simultaneously (Vaughan et al. 2000). This may be a means to keep tracking of a predator in high vegetation and may signal to the predator that it has been spotted. However, stotting by Thompsons's gazelle (*Eudorcas thomsoni*) also has been noted in intraspecific contexts, so its function actually is unknown (Walther 1969; Goodenough et al. 2001).

Most animals are not solitary, so we would expect group mechanisms to evolve (Bertram 1978). As mentioned earlier, living in a group adds more "eyes" to detect a predator, thereby lowering the probability of a given animal in that group to being preyed upon. Mule deer may emit an olfactory signal via an enlarged metatarsal gland at alarm. Injured minnows (and probably many other fish), whose skin has been broken, produce an alarm substance known as Schreckstoff from the skin cells; conspecifics that detect this substance hide and reduce activity (Goodenough et al. 2001). Toxins produced by tadpoles of toads (*Bufo* spp.) act as an alarm substance when predators, such as dragonfly (order Odonata) naiads, are present; "bufotoxin" in the skin of adult toads and larval toads are likely responsible for this.

Some species have sentinel roles (act as guards or lookouts), who sit in an exposed place to search for approaching predators; often relieved of duties; if a predator is seen, the sentinel sounds an alarm (Bertram 1978). Sentinels are found in several species, including dwarf mongoose (*Helogale undulate*) and meerkat (*Suricata suricatta*) (Clutton-Brock et al. 1999). Sentinel individuals also will serve others, provided that they have the same predator, as in Thompson's gazelles and Grant's gazelles (*Nanger granti*) at the Serengeti National Park of Tanzania in response to a common predator, the cheetah (*Acinonyx jubatus*).

If a social mammal is part of a group, there likely is a lower probability of being the one preyed upon (Bertram 1978). This dilution effect seems to work, if a predator that encounters a large group is as likely as encountering a single individual or small group and if there is a limit to the number of prey killed per encounter with a predator. However, the idea that there is safety in numbers instead may increase predation rates because predators may aggregate where prey is abundant.

If an animal is social and lives in a group, it may find that being positioned in the central part of a group is the safest (Bertram 1978). Male northern male elephant seals (*Mirounga angustirostris*) are safer from killer whale (*Orcinus orca*) if located in groups on shore but located away from shoreline. Perhaps this is why tadpoles and some fish form groups when alarmed. However, an important question in the formation of a group or school is whether the central location always the safest? (Parrish 1989). An example is silverside fish (*Menidia menidia*), which lives in groups. A group of silverside fish may be split by predatory seabass (*Centropristis striata*) that swim into the center of the silverside school.

Animals that live in groups also may confuse a predator. Small birds in a flock may sit motionless in foliage at the approach of a hawk (Morse 1980). All of the birds may give alarm calls, which are hard to localize and distract hawk from any particular individual bird in the flock, e.g., the more the attention is divided, the greater the possibility of failure by the predator in locating its prey. Schooling or other social groups may result in a Trafalgar effect, which is named after the battle signals of Lord Nelson's fleet at the battle of Trafalgar (Caro 2005). When animals are in tight groups, individuals can spread information about the location of a predator more quickly. Schooling allows fish to "explode" in all directions. Hence, predators of fish may be more successful if they restrict their attacks on individuals that have strayed from the school or have a conspicuous appearance. As mentioned earlier in Sect. 7.1, stripes in zebras may confuse a predator.

Chapter 8
Habitat Selection

8.1 Selection Versus Use of a Substrate or Habitat

Habitat selection of a substrate or habitat to minimize predation implies a behavioral choice of the appropriate background. Use of a habitat may not imply a behavioral choice. Selection is a largely unexplored area, but it seems that some species select a background to rest in which it makes these animals less conspicuous to predators.

Commensal mice (family Muridae) in the USA and throughout the world have long been of interest. "Commensal" means sharing the table (resources) with humans. There are three classic commensal mice now found worldwide: house mouse (*Mus musculus*), black rat (*Rattus rattus*), and Norway rat (*R. norvegicus*) (Jackson 1982). House mice and black rats arrived in the USA in the seventeenth and early eighteenth century on ships with colonists, whereas Norway rats came to the USA a bit later during the American revolution.

Before Norway rats arrived and became established in the USA, black rats flourished (Jackson 1982). Because the Norway rat presumably is much more aggressive than black rats, Norway rats are believed to have displaced black rats from most parts of the USA; black rats common in more southerly areas, as in California. Interference competition between two species will be discussed later in Chap. 18.

Northern flying squirrels (*Glaucomys sabrinus*) and southern flying squirrels (*G. volans*) are common in North America, but the former seems to be restricted to western and northern North America (Yahner 2001). The range of the northern flying squirrel also is relatively discontinuous in the eastern Appalachian Mountains; it is a federally endangered species and occurs at high elevations, particularly in isolated high-elevation areas of the East. Today, it occurs as "population pockets" once the glaciers receded about 11,000 years ago in the Pleistocene. Thus, habitat loss for the northern flying squirrel in the Appalachian Mountains is a critical factor in the conservation of this species. Its close relative, the southern flying squirrel, has a parasitic nematode (*Strongyloides* spp.) that is harmless to the southern flying squirrel but usually is fatal to the northern flying squirrel. Hence, parasites and habitat selection can limit the abundance and the distribution of northern flying squirrels.

8.2 Testing Habitat Selection Versus Use

Which is more testable—habitat use or habitat selection? Is one of these two concepts easier to test in the lab or field? In my opinion, habitat selection is more testable than habitat use, and it is probably easier in the lab than in the field. A start is to show correlations with environmental variables and to statistically test, for example, with chi-square tests. But even if show habitat selection in the lab, is this what really happens in the field, or are the variables measured or manipulated in the lab relevant to field populations? These questions always have to be asked in studies of wildlife behavior.

Moreover, do correlations between environmental variables and behavior show causation? Again, in my view, correlations are a start, particularly if have no clue as the habitat requirements or general natural history of an organism. For instance, we know much about the natural history and behavior of birds and mammals but not with other taxonomic groups, such as amphibians (Yahner 2004a).

In field exclosures, habitat selection has been invoked in meadow voles (*Microtus pennsylvanicus*) and montane voles (*Microtus montanus*) (Douglas 1976). But even if we can show, or suspect that, habitat selection has occurred in a given species, we have to ask whether this habitat selection will remain the same over time or in given habitats. For instance, some wildlife species may have adapted to human environments, e.g., white-tailed deer, in urban areas.

8.3 Habitat Selection and Urban Wildlife

Populations of white-tailed deer have increased dramatically over the past few decades (Yahner 2000). These deer are now in every state and probably number around 20 million nationwide. At Valley Forge National Park, which is a national park in southeastern Pennsylvania and is surrounded by residential areas, the major mortality factor for deer is collisions with cars. In 1993, costs per collision estimated at $1,600 per collision and cause about a 4% injury rate and 0.03% mortality rate in humans. Suburban dwellers near and adjacent to Valley Forge cannot grow shrubs in yards or have gardens without deer eating these plants. Therefore, habitat selection by deer today in urban areas presumably is much different than when Washington and his troops crossed the Delaware and lived in winter encampments at Valley Forge in the eighteenth century (Cypher et al. 1988).

Deer impacts on farm crops are probably very severe (Vecellio et al. 1994). In Virginia, 55% of crop producers experienced severe damage by deer, and 26% of homeowners experienced severe damage to shrubs by deer. Certainly, deer have adapted to new food resources and to presence of humans.

Gray squirrels often are associated with urban areas (Edwards et al. 2003), and this squirrel may become adapted to humans (Cooper et al. 2008). For example, alert distances of gray squirrels with two levels of human activity (low = no sidewalks or open areas versus high = few trees, sidewalks, human traffic) was greater in low-use areas (10 m) compared to high-use areas (5 m). When two different stimuli (human only or human with a 6-month-old golden retriever on a leash) were tested in these two areas of human use, the presence of a dog affected alert distances only in high-use areas (6 m in absence versus 4 m in presence of a dog). Hence, animals in urban environment may adjust their behavior to the presence of humans.

8.4 Habitat Selection and Wildlife Recolonization

Many animals, like gray wolves, are returning to their native range (Treves et al. 2002). In Wisconsin, gray wolves have recolonized the state since the mid-1970s and now number around 250–300; as a result, farmers have complained about wolves preying on domestic animals. However, can wolves be really blamed for this predatory behavior associated with recolonization of areas? Wolves cannot read signs saying "don't eat domestic animals." Between 1976 and 2000, for instance, The Wisconsin Department of Natural Resources received 176 complaints about domestic animal loss to wolves, with 49% verified as being wolf predation, as based on tracks, scat, and bite marks. As a consequence of this wolf predation, greater than $150,000 was paid as compensation to these farmers. The Wisconsin Department of Natural Resources is considering using a South American camel, called llamas (*Lama glama*), or large domestic dogs, as guards of domestic animals. Furthermore, in the northern part of the Greater Yellowstone Ecosystem (GYE), concern has been expressed about declines in elk populations (e.g., a decline of 33 calves per 100 females to 13 calves per 100 females from 2003 to 2006; 31 gray wolves have been introduced in 1995–1996 in the GYE. Numbers of black bear have increased (tripled) since the 1990s in the GYE; in summer, loss of elk calves was 60% in northern GYE due to predation by black bear and 17% by wolves; in addition to these two major predators, three other predators on elk calves occur in the GYE (brown bear, mountain lion, and coyote). Elk calves in the GYE had low values of gamma globulins in their blood; low values of gamma globulins have been noted elsewhere in white-tailed deer that were susceptible to both disease and high mortality. Implications of this research are to reduce human harvest of elk (more conservative of the two implications) or control both bear and wolf populations in areas near GYE.

8.5 Some Factors Affecting Habitat Selection

What about humans or their activities affecting habitat selection in wildlife? Brown bears will tolerate humans at close range under certain circumstances (Herrero et al. 1985). Brown bears, and probably other wildlife, may conserve energy by habituating to humans. In the case of brown bears, they "learn" that humans along roads, for instance, are "harmless;" in Yellowstone National Park, bears may not overtly react to people, who are as close as 20–50 m; in some other areas (e.g., along some coastal areas), this overt reaction distance may be less than a few meters. Thus, reaction to humans by dangerous wildlife may vary regionally. Perhaps this regional difference in reaction distance by wildlife to humans may be related to resource distribution. If food resources are clumped, as with salmon (family Salmonidae) along rivers, brown bears will tolerate other bears and humans at close range, which has led to the development of bear-viewing sites along rivers.

With songbirds, rates of song may vary with habitat size and species (McShea and Rappole 1997). For instance, two forest birds, wood thrush (*Hylocichla mustelina*) and ovenbird (*Seiurus aurocapillus*) sing more often in larger forest tracts (greater than 50 ha) than in smaller forest tracts (4–24 ha); yet, in an early successional species, the northern cardinal (*Cardinalis cardinalis*) song rates were the reverse, being lower in larger than in smaller forest tracts. In other studies, not all male ovenbirds were mated; instead, the chance of a male being mated was directly related to forest tract of a given size, not all males are mated, affecting the estimation of population numbers and affecting what we might term "habitat quality"—in large forest tracts (>500 ha), 76% of male ovenbirds paired, whereas in small tracts (9–150 ha), only 25% of birds paired (Gibbs and Faaborg 1990; Villard et al. 1993), but pairing success of wood thrushes did not vary among woodlots ranging from 3 to 12 ha (Friesen et al. 1999). Because wildlife conservationists rely on song rates to estimate bird abundance (Sauer et al. 2008), and each singing male is believed to be paired, estimates of abundance, and, hence, data interpretations, may be influenced by the size of forest tracts in a given landscape.

Habitat selection by nonbirds or nonmammals may be difficult to determine because we may be looking at the wrong variables or the variables may be difficult to measure. In insects and salamanders, for instance, cues used by these taxa in habitat selection may be weather or climatic variables. Both taxa are relatively scarce along forest edges (Yahner 1988). In warm mornings during spring, insects presumably abundant in sun-lit areas along edges (Bramble et al. 1992). Perhaps butterflies select agricultural landscapes more so than forested or residential landscapes because of interaction between microclimate and food (Yahner 1999). Perhaps the degree of openness plus availability of wildflowers as a nectar source are why butterflies much more common in agricultural landscapes than in forest or residential landscapes.

At forest–farmland interfaces in central Pennsylvania, only 11% of woodland salamanders were found within 5 m of the forest edge (Young and Yahner 2003); these edges were much dryer because of light penetration along the edge. Also, in managed stands, no woodland salamanders were found; e.g., redback salamanders (*Plethodon cinereus*); instead, this salamander was found where abundant trees, as in uncut stands where light penetration presumably did not dry out the forest floor (Rodewald and Yahner 1999).

Some animals, like least weasel, ermine, and long-tailed weasel, may select habitat on the basis of snow cover (Yahner 2001). The initial two species are smaller and occur in more northerly latitudes, whereas the long-tailed weasel is larger and occurs in more southerly latitudes. The long-tailed weasel cannot exploit the subnivean environment, which is the zone between the ground level and snow cover, where the small animals, which are principal food resources, are active throughout winter. Another variable limiting habitat selection and distribution of the long-tailed weasel in northern latitudes is predation on the long-tailed weasel by the arboreal weasel, termed the American marten (*Martes americana*).

8.6 Other Factors Affecting Habitat Selection

8.6.1 Ambient Temperature

Habitat selection by eastern cottontails (*Sylvilagus floridanus*) may vary as a function of microhabitat (Swihart and Yahner 1982; Althoff et al. 1997). Throughout the year, cottontails typically use old-fields and shrublands; but in summer and fall when crops mature, cottontails use these habitats. In winter, shrubland and woodland habitat is important for cover. Daytime resting (bedding) sites used by eastern cottontails tend to be found in dense cover, which likely reduces predation risks and minimizes heat loss. For instance, in winter, female cottontails used burrows when ambient temperatures dip below 0°C.

A question that needs to be asked is "Can global climate change affect habitat selection?" (Yahner 2000). In 2001, the National Academy of Sciences confirmed that temperatures are rising (Bolen and Robinson 1995). If the polar bear (*Ursus maritimus*), which is viewed by many as a charismatic megavertebrate, is an icon of global climate change. Ice is melting 3 weeks earlier per year than it did 30 years ago. The breaking up of pack ice via melting is affecting the time available to polar bears to hunt pups of ring seals (*Pusa hispida*). If food in the form of ring seals becomes limiting, these bears may move ashore, causing a greater chance of encounters with humans.

Sedentary species are less capable of moving to new habitats compared to more mobile species (Yahner 2000). A concern among conservationists is where will species adapted to high latitudes or elevations escape if global climate change proceeds. Furthermore, species either with low numbers to begin with or restricted geographic range, e.g., endangered species, may be most susceptible to changing habitat conditions brought upon by climate change compared to species with high population numbers or a wide range. Warmer winters can increase the abundance of forest insects, e.g., caterpillars (order Lepidoptera), which may have a detrimental effect on forest trees via defoliation; as a consequence, birds that feed on these caterpillars, such as black-throated blue warblers (*Dendroica caerulescens*), may increase in abundance and distribution. At least 50 species of songbirds may increase their winter distribution northward; it has been found that in North American birds, species averaged a shift of 34 miles northward over past 26 years; in the blue–gray gnatcatcher (*Polioptila caerulea*), the shift northward has been 314 miles; similar northward shifts have been noted in British birds (Hitch and Leberg 2007).

Changes in global climate temperatures may have other impacts on wildlife (Yahner 2000). A projected 3°C increase may affect the loss of 9–62% of the small mammal populations on 19 isolated mountain ranges in the western USA. In Texas, hibernacula used by wintering bats may be affected because of microclimatic changes in caves, with bats preferring a certain part of the cave, e.g., entrance, as a location of hibernation. In the southeastern USA, the red-cockaded woodpecker (*Picoides borealis*) relies on mature living trees as nest sites for cavities; if global climate change causes mortality of these trees, then it may affect the habitat selection of this woodpecker. Warmer stream temperatures may negatively affect habitat selection by trout (subfamily Salmonidae) but be beneficial to warm-water fish; for instance, with warmer water temperatures, a 29–33% increase in walleye (*Sander vitreus*) production predicted in Lake Michigan. Range expansion into new habitats in the Midwest by common raccoons may be attributed to an increase in mean ambient temperatures (Larivière 2004). In the case of the common raccoon, longer growing seasons and earlier-maturing crop varieties, combined with warmer winters may decrease denning period and reduce fat reserve needs during winters. There also is concern about the distribution of caribou in Arctic regions because of warmer winters, effects on vegetation, and increased insect harassment (Grayson and Delpech 2005), as evidenced by glacial receding during the Pleistocene, which may have caused the extinction of caribou from southern France.

8.6.2 Acid Deposition

Acidity resulting from acid deposition (rain, snow, and particulates) can change the chemistry of streams, lakes, and ponds, thereby affecting the suitability of these habitats to fish and aquatic insects (Yahner 2000). In the 1980s, 38 (46%) of 61 headwater streams in southwestern adults in streams with a pH of 4.2; spotted salamander (Fig. 8.1) (*Ambtoma maculatum*) occasionally survived in these same streams, but adults were undersized and probably did not reproduce well.

In forest soils of New York, lower densities and number of invertebrate species were noted in soils with high acidity; and in Europe, limited evidence found that snails (class Gastropoda) in acidic soils were calcium deficient (Yahner 2000). Great tits (*Parus major*) that fed on these snails had eggs with thinner shells. Thus, environmental changes or contaminants can affect the "selection" or "preference" indirectly of certain organisms.

8.6.3 Population Density

It is often difficult to show is population density can affect habitat selection, but if one of two habitats is preferred and the other is less-preferred, a population theoretically may use only the preferred habitat (Partridge 1978). For example, in Alaska, populations of brown lemmings (*Lemmus sibiricus*) are very cyclic; researchers have shown that a narrower range of habitats is used by lemmings when densities are low; in years with peak numbers, lemmings are found in all terrestrial habitats.

8.6 Other Factors Affecting Habitat Selection

Fig. 8.1 The spotted salamander of eastern North America. It is considered as a mole salamander

Perhaps population density is a factor in habitat selection, depending on the species, intraspecific competition for space, home sites, food, or other resources (Partridge 1978). But because a certain species may have high population density in a given habitat and occurs in lower density, we simply cannot conclude that this is a "spilling over" effect of too many animals. If population density is high in a seemingly preferred habitat, may be better for the animal to settle or choose a less-preferred habitat; hence, a new arrival, for instance, may be better off settling in a less-preferred and less-crowded habitat (this was proposed as early as in the 1970s by Fretwell and Lucas).

This distribution of animals among habitats based on their choice has been termed as the Ideal Free Distribution Model (Partridge 1978). For example, if food is twice as abundant in habitat A compared to habitat B, we might predict that twice as many animals might occur in habitat A. But it might be better to stay in a less-preferred habitat as a helper; then when older kin die, inherit the habitat (Goodenough et al. 2001).

Possibly another factor acting as a cue in selecting a habitat may be social stimulation (Morse 1980). In some ways, this may be the opposite of high or low population density acting as a factor in habitat selection. The fact that other individuals of the same or different species can affect habitat selection is a process known as social stimulation. In oysters (class Bivalvia), metabolic products of conspecifics may be detected by larvae via chemoreceptors, which simulate other oysters settling in an area. The presence of territorial, singing male songbirds may indicate to other males of the same species that a given habitat is suitable (Morse 1980).

8.6.4 Tradition

No discussion of factors affecting habitat selection would be complete without mention of tradition (Morse 1980). From a wildlife-behavior perspective, tradition may be defined as an attachment to a site from one generation to another. Typically, young leave their birth site (natal site) and disperse from the mother social unit, but female prairie dogs (*Cynomys* spp.) leaves the coterie (social unit) to her young and move to the periphery of the colony. Similarly, a queen honeybee (*Apis* spp.) leaves the hives to her daughter. In these species, leaving a habitat known to be suitable to young, presumably is adaptive to the young.

8.7 Is Habitat Selection Learned?

A difficult question to answer is whether habitat selection is learned because of the relative importance of innate (genetic) and learned elements in habitat selection is unknown (Partridge 1978). With urban wildlife, there is probably a lot of learning by young animals. For instance, pigeons or rock doves (*Columba livia*) presumably has learned that window ledges and bridges serve as nesting habitats like cliffs in its natural range (Bolen and Robinson 1995) before pesticides decimated populations, peregrine falcons (*Falco peregrinus*), also are cliff nesters in their natural range, regularly nest on window ledges in city skyscrapers, especially where pigeons, which are major prey, are abundant. As a consequence of habitat selection and breeding success of peregrines in cities, peregrines have been delisted from the endangered species list in 1999.

Natal site may not be important to nest-site selection in some species, e.g., Cooper's hawks (*Accipiter cooperii*). In urban areas, this hawk uses the general structure of nest sites in selecting a habitat; a grove of large trees, not an urban area, is selected; thus, nest-site selection in Cooper's hawks seems to be independent of natal experience (Mannan et al. 2007).

Can habitat selection be shown to be adaptive (Partridge 1978)? In other words, is habitat selection a random process or can it be demonstrated that animals in certain habitats selected habitats to maximize survival and reproduction? As we have seen, there is differential fitness in fishes and amphibians in acidic waters. Thus, this thinking has led researchers to examine differences between habitat sources and sinks. Large habitats, for instance, may act as sources (preferred habitats) of individuals that may occupy sinks (nonpreferred habitats). Although sinks may contain individuals, these individuals may not breed and produce young.

Habitat selection in most animals may be limited by energy and time available to search for a suitable habitat, with time and energy better devoted to reproduction, finding food, or reducing susceptibility to predation (Morse 1980). Many organisms may have limited ability to find suitable habitat because of dispersal abilities or because they may live only for a short time (Partridge 1980). Moreover, some organisms may have limited ability to find suitable habitat because genetics determines broad habitat selection, but the details as to which habitat is better may be fine-tuned after an animal has sampled various habitats, e.g., learning involved later.

8.8 Role of Resources: Generalist Versus Specialist

Considerable work has been done on what constitutes a habitat generalist compared to a habitat specialist (Partridge 1978). To place arbitrarily an animal in one or the other category, but to know why an animal choose a given habitat requires much more detailed syntheses of resources available and those used. This dichotomy between generalist and specialist is very subjective because we may not know what an animal may be focusing upon in selecting a habitat with a given set resources or we may not know how these resources affect fitness of an animal making this choice.

Typically, a generalist uses a breadth, or range, of habitats compared to a specialist that is very specific in its habitat selection (Partridge 1978). However, animals can become specialists on food resources as perhaps to minimize competition, as with two aquatic mustelids, the northern river otter (*Lontra canadensis*) (Fig. 8.2) and the mink (*Mustela vison*) (Yahner 2001). Conversely, unlike the otter and the mink, some animals may be generalists in food selection but specialists in home-site selection. For instance, consider the black bear. This bear is an omnivore, which means it eats a broad range of plant and animal food items. Black bears when it comes to home sites (den sites) for winter lethargy, it becomes a specialist. In southern latitudes, black bears tend to select aboveground den sites, in which tree cavities averaging 11 m aboveground. Winter lethargy in tree cavities protects bears from winter rains, dogs, etc., because no insulating snow cover. In northern latitudes black bears almost always use dens at or below (about 1 m) ground level, as under uprooted trees. This is an example of habitat selection changing regionally within the same species. In another den-site selection study by black bear (White et al. 2001), tree dens were used exclusively by females, where trees were not extensively harvested. In a nearby site, trees were harvested, but this site had significantly more relief; females in this area nested on the ground, suggesting that because of tree removal, cavities (resource) availability was altered. Apparently, older females (in particular) apparently "learned" to use elevated sites as wintering dens when tree cavities were at a premium.

As with black bear, common raccoons in agricultural areas of southern latitudes also used tree dens (Henner et al. 2004). Females tend to use tree cavities, whereas males use ground dens or brush piles; in addition, females select sites near cornfields, but males select sites near water. Thus, habitat selection can be important in den-site selection. By knowing this, tree removal can affect female raccoon near croplands or be a detriment as a management strategy.

Fig. 8.2 The northern river otter, or North American river otter, is found only in (endemic) North America. It is a weasel, which is well-adapted to aquatic habitats

The habitat generalist versus specialist dichotomy may simply be a function of resource availability. In Minnesota shelterbelts, which are human-made habitats and lack cavities, I erected 22 nest boxes (artificial tree cavities) spaced at 50-m intervals (Yahner 1984). Once these boxes were erected, one black-capped chickadee and 20 house wrens (*Troglodytes aedon*) used them within a 2-year period. Neither species nested in shelterbelts over a 3-year period prior to putting up these boxes (Yahner 1983). In another study, 50 nest boxes (designed for eastern screech owls, *Megascops asio*) were placed in each of three habitats: agriculture, suburban, and woodland (Fowler and Dimmick 1983). Eastern screech owls, gray squirrels, screech owls, and southern flying squirrels (*G. volans*) used boxes at each of the sites in winter and in spring, whereas European starlings (*Sturnus vulgaris*) used boxes only in spring to breed. Over a 3-year period, 48 of 50 boxes in agriculture, 49 of 50 in suburbia, and <50% in woodland were used, which suggest that cavities are scarcer in woodland, not that habitat selection varies with habitat. In breeding areas of Barrow's goldeneyes (*Bucephala islandica*), erecting nest boxes can be very important because of intense logging pressure and because Barrow's goldeneye is a species of special concern.

The very common downy woodpecker (*Picoides pubescens*) is a primary cavity nester, e.g., excavates its own cavity. When artificial snags, e.g., dead trees made of out of polystyrene painted brown, are available to these woodpeckers, females avoided the taller snags and placed cavity farther from the top compared to males. This study suggested that provision of artificial snags may be used to restore downy woodpeckers on lands lacking snags (Grubb 1982).

Songbird boxes have been used commonly to restore species. The eastern bluebird (*Sialia sialis*) is a perfect example; it was once common in agricultural areas throughout the USA when wooden posts were used for fences; but when wooden posts were replaced by metal posts, places for cavities became a real premium, causing this bird to decline. Now, with boxes being erected along Bluebird Trails, this bird has made a major comeback (Yahner 2000).

Cavities also may be used to manage mammalian species. For instance, northern flying squirrels may be limited by den sites and food (truffles) (Carey et al. 1997). These squirrels use cavities in live trees and in dead old-growth trees; some create stick nests in secondary-growth trees. From a management perspective, the preservation of cavity trees and dens may be fruitful in managing for northern flying squirrels, which may be beneficial in the recovery of threatened northern spotted owl (*Strix occidentalis*); northern flying squirrels are major prey of owls.

Each animal must have adaptations to be a generalist or specialist. For example, the teeth of reptiles (class Reptilia) are basically designed to allow the animal to feed on any animal, depending on it body size (Vaughan et al. 2000). In other word, teeth of reptiles are jack-of-all-trades. Reptiles do not have deciduous teeth like a mammal (class Mammalia) adapted to lactation.

Being a generalist or a specialist may change over time, seasonally, spatially, or with sex class, e.g., young American alligators (*Alligator mississippiensis*), which specialize on insects when younger, but on vertebrates when adults (Zug et al. 2001). Some animals have other specialized body parts to exploit certain specific resources;

for instance, brown bear have a large muscular back and long claws to dig up burrowing food, e.g., ground squirrels (*Spermophilus* spp.) (Yahner 2001). As a consequence of these adaptations, there may be trade-offs—brown bears cannot climb as adults, but black bears can. Adaptations may also predispose an animal to specific habitats or resources within a habitat, giving each animal a unique role in the habitat (Bolen and Robinson 1995).

8.9 Some Problems in Quantifying Habitat Selection

Habitat selection, or any behavioral work, is difficult to quantify because they may be very little known about the life history, as, for instance, in some amphibians (Yahner 1997b). If want to know something about habitat selection or the impact of habitat change or loss; these databases are not available for certain taxa that have long-term life spans (Yahner 2004b). Many tropical animals are poorly known, let alone described; if their habitat distribution is limited, it becomes critical to determine habitat selection factors. Habitat selection can vary temporally, spatially, and demographically.

Habitat selection may be difficult to determine for wide-ranging species, e.g., migratory songbirds. Since about the late 1960s, radio telemetry has been used to monitor movements of free-, wide-ranging animals. First, an animal need to be captured and fitted with radio telemetry equipment. This capture and marking can be costly, time consuming, and requires expertise and practice. Often, when using radio telemetry, we may not see the animal, and, hence, question the quality of the locations obtained by telemetry. For example, we can never be sure that the presence of the investigator is not affecting the location or the behavior of the animal being monitored.

Habitat selection is not always determined by radio telemetry, but other remote methods can be, and have been, used, including satellite imagery, aerial photographs, and topographic maps. Some techniques that seem logical at first, e.g., counts of pellet groups to assess deer density, are no longer used for these purposes, but they still may have value in estimating habitat selection. In other words, if a pellet group is found in a given habitat, then presumably it is used by at least one deer. Hence, measuring habitat selection requires some skills, consensus on how to best do it, common sense (labor and monies required), and how to interpret the findings.

Chapter 9
Home Range and Homing

9.1 Some Comments About Home Range

We have dealt with dispersion and habitat selection, so how does home range relate to dispersion patterns and habitat selection? A home range might be defined simply as an area in which an animal "normally" lives and which contains many essential requirement, e.g., food, cover, and water (Brown 1973). Further, a home range is exclusive of migrations, emigrations, or erratic wanderings and is specified without reference to presence or absence of aggressive behavior. If an animal establishes a home range in a given habitat, the animal is using that habitat, so has the habitat selection taken place? This is an important question to ask with regard to home ranges, as is how large is a home range and does its use or size vary seasonally, with sex class, or with age class, etc.? Moreover, what factors affect placement or size of home ranges? Is there a distinction among home range, core areas, and territories, or are these interchangeable terms? As we shall see, there are many other questions regarding home range.

9.2 Relationship Between Home Range and Dispersion Patterns

Recall that animals seldom occur randomly in the habitat; they occur either in a regular or in a clumped dispersion pattern, as true of most mobile animals. However, when we speak of dispersion patterns, we almost imply that animals are fixed in space, which is certainly true of some sessile animals, e.g., barnacles, and of plants (Brown 1973). Thus, when dealing with mobile animals, must assess home range via repeated sampling over time because of their mobility.

9.3 Homing in Wildlife

Before dealing with the concept of home range, we need to mention that home range is very different from homing. Homing might be defined as the ability to navigate. These abilities have been made famous by homing pigeons, which have true navigational skills (JWM 2001, p. 1300). Wildlifers have attempted to place animals in back into their native habitat (termed translocation) or to remove nuisance animals from problem areas. But simply moving an animal from one place to another does not mean that the animal is "saved" or that it will not home to its original place of removal (JWM 2001, p. 1300). For instance, red-legged frogs (*Rana draytonii*) can home at least 500 m between ponds if translocated. This homing information is important because this frog is native to California, has a threatened status, and is found in only 25% of historic range. So, to protect these frogs from human disturbances, translocations have been used.

Homing tends to be better developed in animal, e.g., mammals, with extensive home range. Homing has been examined in nuisance black bears. As a means of getting information on this phenomenon, Rogers (1986) examined how far a nuisance black bear must be translocated to minimize chances of their returning and again becoming a problem. Based on about 180 adult bears, which were translocated in 11 states and provinces, all successfully homed if translocated within 8–20 km of original capture site. But if distances of translocation were increased to 220 km, only 20% successfully homed. Rogers (1986) also noted that homing ability did not vary between adult males and females. He found no impact of translocation on mortality, but there were lower percentages of homing in bears translocated from Yosemite and Great Smoky Mountain national parks, with poaching outside park boundaries being blamed for higher mortality rates.

Nuisance animals are not limited to black bears. Common raccoons (Fig. 9.1) also can become nuisance animals and are commonly translocated (Mosillo et al. 1999). Translocated raccoons often move considerable distances after release, particularly urban raccoons (up to 60 km); but most released animals moved to areas with human residences and den there, whereas resident animals tended to den in

Fig. 9.1 The common raccoon is native to North America. Habitats of the common raccoon include deciduous and mixed forests, but raccoons are now found in urban and mountainous, and coastal areas. The common raccoon was introduced into Europe and Asia (Japan)

woodlands. Thus, translocated animals can add to problems with the spread of disease, e.g., rabies (*Lyssavirus* spp.). This raises an ethical question in that should nuisance species, e.g., common raccoons, be translocated or euthanized?

9.4 Quantifying a Home Range

There probably are as many ways to measure home range as there are definitions of home range (Fuller et al. 2005). Probably, the earliest method used to measure home range is to establish an observation–area curve, whereby the location of an animal is plotted at specified time periods. Sometimes, this approach is called the minimum convex polygon method. Some methods of for the measurement of the size or the extent of a home range can be very sophisticated methods.

With mobile animals requiring sophisticated methodology to determine locations, e.g., radio telemetry, measurement of locations every 5–10 s may be prohibitive or unreasonable. Thus, it may be that monitoring an animal less frequently per night may give similar results (Harrison 2002).

9.5 How Large Is a Home Range

Many factors seemingly affect the interspecific size of a home range. For instance, we would expect that larger, more-mobile animals have the largest home range; shown in many studies; found in mammals, but not so in birds and reptiles, that body weight and home range size are directly related (Brown 1973). Size of home range size also varies with foraging strategy. Size it typically larger in predators than in grazing herbivores. Grass, as a food resource for grazing herbivores, is hypothetically less sparsely distributed than are prey for predators.

Size of a home range also can vary intraspecifically with sex class, age class, or time of season (Brown 1973). Hence, a more reasonable definition of home range may be "the extent of an area with a defined probability of occurrence of an animal during a specified time period." Size of home range can vary markedly in adult males during the breeding season than during the nonbreeding season (Yahner 1978d).

In mountain lions, male home ranges overlap those of several females but seldom overlap those of other males (Feldhamer et al. 750). Based on numerous studies of mountain lion, home ranges of males is greater than those of females. In another mobile carnivore, gray fox (*Urocyon cinereoargenteus*) home ranges were similar between adult males and adult females, but they did vary seasonally (Chamberlain and Leopold 2000).

Size of home ranges of adult common raccoons, a medium-sized omnivore, were similar between adult males and adult females, but size varied with the type of habitat [forests (mean), 254 ha; agriculture, 137 ha; urban, 89 ha] (Compton 2007). In this same study, size of home range also differed with season; in forest, the mean size of home ranges was 245 ha in spring and summer compared to 178 ha in fall and winter.

In male white-tailed deer, which represent a large-sized herbivore, size of home range declined 56% with age (Webb et al. 2007). Size averaged 416 ha in yearlings (1.5 years old), but it declined to 182 ha in the same deer when they became 5.5 years old. In adult male Appalachian cottontail (*Sylvilagus obscurus*), a small-sized herbivore, size of home ranges was 5.7–13.3 ha during leaf-on season (May–September) but only 1.5–9.0 ha in leaf-off season (October–April) (Northeastern Nat. 2007, p. 99). During the leaf-on season, food and cover were abundant, but during the leaf-off season, these resources were relatively less common to Appalachian cottontails.

Bachman's sparrow is a species of concern, which is common in southern pine (*Pinus* spp.) plantations and mature pine stands (Stober and Krementz 2006). In this bird, average size of home range were similar (2.95 ha) in males and females, but size in both sexes was higher in mature stands (4.79 ha) than in plantations (<3 ha).

9.6 What Is the Shape of a Home Range?

Home ranges may not be circular or elliptical but may vary with physical features of the habitat or of the way-of-life of an animal (Fuller et al. 2005). For example, northern river otters are aquatic animals that live in and sometimes right along the shoreline of rivers or lakes (Serfass et al. 1993). Thus, we would expect the shape of home ranges of northern river otters to be dictated by water drainage patterns. Small mammals, e.g., cottontails in intensive agricultural areas, follow linear features, like fencerows; hence, home ranges of cottontails (Fig. 9.2) in intensive agricultural areas are linear as well (Swihart and Yahner 1982).

Often times, all areas of a home range are not used equally, but a portion is used disproportionately more than other areas. This is known as a core area (Brown 1973). A core area is not necessarily a defended area, but instead is defined by animal usage or methodology of a researcher. In gray fox, a home range of males in the nonbreeding season may average less than 200 ha, but a core area may only be about one-tenth of this size (Chamberlain and Leopold 2000). We expect core areas to be located around nests, burrows, or favored feeding area or resting areas. For example, in adult eastern chipmunks, a core area is an area about 15 m in radius around a burrow system (Yahner 1978b).

An understanding of the size of home ranges can be useful under two circumstances. First, in understanding interspecific interactions and impact of an invasion of a species on a resident species, and, second, in relation to disease issues. In the first circumstance, an example is shown by coyote and red fox (Gosselink et al. 2003). Coyotes have expanded their range to include most of North America, but they actually were found in much of eastern North America until about 1,000 years ago, which was well before arrival of Europeans (Yahner 2001). Reasons for the demise of the coyote 10 centuries ago from the East are unclear. However, in the early nineteenth century, coyotes gradually returned to eastern North America from the West, perhaps because Midwestern prairies were converted to agriculture to

9.6 What Is the Shape of a Home Range?

Fig. 9.2 The eastern cottontail is among about 16 species in North and South America. It closely resembles the wild European rabbit

which coyotes readily adapted. By the middle of the twentieth century, coyotes were found in eastern USA and eastern Canada, and today are in every state and province in the East. A second reason given for the range expansion of coyotes into the East has been the extirpation of gray wolves from this area. In addition, some coyotes may have been released into the wild in the Southeast by hunters. As a consequence of this range expansion by coyotes in the East, populations of red fox have declined. This issue of range expansion by the coyote at the expense of red fox will be discussed in depth later when dealing with the concept of competition (see Chap. 18).

Perhaps the "quality" of a habitat affected range expansion of the coyote (Gosselink et al. 2003). For example, in an Illinois study, home ranges of coyotes occurred in over-rich habitats (grasslands, waterways, and no-till corn). On the other hand, red foxes were found in human-associated habitats (abandoned and occupied farmsteads, and rural residential areas). In winter, home ranges of both resident red foxes and coyotes increased nearly fourfold and twofold compared to summer.

A second circumstance in which an understanding of the size of home ranges can be useful is in disease, as with West Nile virus (WNV) and home ranges of birds (Yaremych et al. 2004; Rohnke and Yahner 2008). WNV is a mosquito-borne virus that was first detected in New York, USA in 1999; now it is in most of the continental USA, seven Canadian provinces, throughout Mexico, and parts of the Caribbean; prior to this, it was found in Africa, eastern Asia, and the Middle East. This virus occurs in at least 48 species of mosquitoes (but we must understand that certain species are bird-specific, mammal-specific, or generalist feeders), >250 species of birds, in at least 18 mammal species, and in one crocodile and one alligator species. According to the Center for Disease Control, greater than 15,000 people in the USA affected from 1999 to 2005, with more than 500 human deaths; however, most humans experience mild or no symptoms if the virus is present, but WNV is 100% fatal in vaccinated domestic horse (*Equus caballus*).

All vertebrates are susceptible to WMV, but only 12 of 740 songbirds of 45 species in central Pennsylvania tested positive for antibodies of WNV (Rohnke 2008). Eight of these were gray catbird (*Dumetella carolinensis*); in addition, over 2,000 specimens of the principal vector of WNV, the urban-dwelling mosquito (*Culex pipiens*), did not contain antibodies of WNV. If home ranges of songbirds overlap residences of humans, then we need to be concerned.

As a presumed adaptation to increased urbanization, abundance of American crows increasing (Yaremych et al. 2004). Because American crows are very susceptible to WNV and if WNV is prevalent when mosquitoes are most abundant (spring–summer), then we need to know something about the home ranges of American crows in various habitats. First, size of home ranges of crows is much smaller in urban areas than in nonurban habitats. Second, home ranges of crows are more likely to occur in urban habitats and agricultural habitats with low- to medium-density of humans than in forested habitats or in high-density urban habitats. Size of home ranges did not differ with sex class or age class. Thus, the fact that crows with a relatively large home range and the fact that his bird selects agricultural habitats for feeding and select urban habitats for roosting at night near humans, this species has the potential importance to transmit WNV to other animals and, hence, transport WNV across a relatively large area.

Chapter 10
Spacing Mechanisms

10.1 Introduction to Territoriality

Dispersion, habitat selection, and location of home ranges are affected by physical and biological factors (e.g., food). Behavior can affect where an animal occurs, but often resource managers have overlooked this factor and the public certainly is not aware of it. For instance, why can white-tailed deer herds in winter achieve densities of 200 deer per square miles, yet we do not see red foxes in large groups during winter? A major reason is that red foxes exhibit territoriality.

What is a territory and how does it differ from a home range? Sometimes these can be the same unit of space, but typically a home range is larger in size than a territory used by a given animal. Most importantly, territoriality involves agonistic behavior or aggression against rivals around a fixed area (Brown 1973). Hence, territoriality involves active efforts to defend space.

A species that defends a territory is termed territorial (Brown 1973). As mentioned, we may define a territory as a fixed, defended area, e.g., it is delineated as a space in the habitat, not by the area around the individual animal. This is important when we speak of other spacing mechanisms, such as the concept of defense of individual space.

A territory simply is not an area of dominance (Brown 1973). Typically, a resident, territorial animal is dominant in its territory against intruders, but in order for territoriality to be invoked, need to not only have aggressive behavior, e.g., actual attack or threat, but also the intruder must be repulsed from the territory by this behavior of the owner (e.g., attack, threat, scent, song, etc.). However, this does not mean that an intruder never enters another territory, but rather, the intruder is driven out. Thus, a territory is not impenetrable, and it is not an exclusive area. As a result, territoriality simply cannot be inferred by a given dispersion pattern. Hence, a tendency to refer to a species with no overlap in home range as being territorial is erroneous. For instance, a North American weasel, the fisher (*Martes pennanti*), has a nonoverlapping home range with a uniform distribution, but it is not considered territorial (Koen et al. 2007). Also, many cats (e.g., mountain lions) maintain exclusive areas and are solitary, which may not necessarily imply that they are territorial (Davies 1978). An area used may consist of pathways from a home site to a feeding area, whereby habitat used radiates out from a home site, e.g., a burrow; so conceivably, there is no territory; simply, it may mean that the species is exhibiting non-overlapping home ranges.

10.2 How is a Territory Defended?

Territoriality is common and best studied in visually oriented species, e.g., many birds, but it has been found in bony fishes, frogs, salamanders, lizards, crocodilians, and mammals (Morse 1980). Defense of a territory, as perhaps in some butterflies (e.g., specked wood butterfly, *Pararge aegeria*), may not involve overt aggression. Rather, the winner of a contest is simply becomes the resident on the territory at the time (Partridge 1978).

However, overt aggression is common in territorial defense and may include threats and actual chases, e.g., eastern chipmunks (Yahner 1978b). In chipmunks, chases are very common when an intruder enters the defended core areas of chipmunks, which is about an area of about 15 m radius around a burrow system; other portions of home range are not defended. The percentage of time spent by a territorial animal, e.g., an eastern chipmunk, often is low overall but considerably lower than actual fighting (1.22 vs. 0.04%, respectively).

Many lizards (order Squamata, genus *Anolis*), which typically are visually oriented and diurnal, use gular displays, that vary in size and color (Pough et al. 2002). These displays are combined with push-ups and head bobs, followed (if necessary) by fighting.

In common chimpanzee (*Pan troglodytes*), males in subgroups patrol boundaries of territories (Pusey et al. 2007). If intruders are encountered by these subgroups, intruders are killed. Overt aggression by subgroups of males toward intruders is second only to disease (respiratory) as a mortality factor. In fact, subgroups with fewer males due to human poaching (for protein sources for humans) make subgroups of males susceptible to other subgroups and less likely to defend a territory.

The northern flicker (*Colaptes auratus*) (Fig. 10.1) uses diving and cavity blocking to defend its nest and territory (Fisher and Wiebe 2006). In flickers, aggression is highest when rearing young, not at nest building. Defense of a nest and a territory does not vary between sex classes nor with size of a clutch.

Song seems to be a major means of territorial defense in many species of birds (Gill 1990). Once a territorial boundary is established, songs likely take the place of overt aggression, being presumably less costly in many ways. Some scientists have suggested that song in birds is used as a long-range warning signal to repel trespassers, whereas visual displays are used to repel trespassers at intermediate distances, with chases and attacks used if an intruder persists in violating territorial boundaries. This use of graded signals may be similar to gular displays followed by other visual displays in territorial lizards when defending a territory.

Tropical birds, called trogons (family Trogonidae), are very secretive and vocalize in mixed-gender calling sessions consisting 3–10 birds during breeding season

10.2 How is a Territory Defended?

Fig. 10.1 The northern flicker is a woodpecker native to most of North America and is a woodpecker that migrates. It uses diving and cavity blocking to defend its nest and territory

(Riehl WJO, 2008, p. 248). But the function of these vocalizations is unclear, but includes enhancing foraging efficiency, helping to select future nest sites, or maintaining territorial boundaries.

Although eastern chipmunks are very vocal, I found no evidence that any of their three major types of vocalizations serve for territorial defense (Yahner 1978e). In coyotes, the group yip-howl acts to space out groups, in (typically urine) indicating that area is occupied (Yahner 2001). In gray wolves, interpack spacing is achieved by a group howl. To understand the role of vocalizations or olfaction in defining the boundaries of a canid, a single scent-mark created by a canid must be visualized (Yahner 2001). A single scent-mark is not directional, as may be a sign interpreted by humans. So, what happens when a lone wolf moves into the center of a territory of another pack? It can be attacked and killed. For instance, 41% of the known deaths of lone wolves occurred by neighboring packs in the 1-km radius border of pack territory overlap; 91% of deaths were within a 3.2-km radius (Mech 1994). Hence, the border of territories and within the territory of another pack is a dangerous place in the life of a lone wolf. Thus, a single scent-mark in canids does not work by itself. Rather, wolf packs (and presumably in other territorial canids) use an olfactory bowl (Mech 1970). The existence of an olfactory bowl in nature often occurs along logging roads, trails, or other established routes within a territory of a pack. Depending on the location if a wolf within its territory, the rate of scent-marking may vary, which can be measured in distances between scent-marks, thereby creating an olfactory bowl. Along the territory boundary, which might overlap that of other packs, a wolf may deposit scent-marks every 110 m; as it goes into the center of the territory, scent-marks are deposited every 180 m. As a consequence, the greater concentration of scent-marks is less along the border than in the center of a territory. A single scent-mark does not serve as a barrier, but rather a gradient of marks is necessary.

The border of a territory of a wolf pack territory may be 10–20 km wide, with the outer 1-km border being shared space (Mech 1994); thus, the outer 1 km acts as an

Fig. 10.2 The beaver, or North American beaver, is the only beaver of North America. It was introduced to South America. The only other living (extant) beaver is the European beaver of northern Europe and Asia

impregnable barrier to neighboring packs or to a lone wolf in the absence of other pack members. Areas of interpack overlap (e.g., territorial borders) perhaps reduce interpack encounters and help lone members of a pack orient themselves for safety.

American beaver (Fig. 10.2) also use scent-marks to delineate territories (Yahner 2001), but do beaver create an olfactory bowl? Beaver are known to use olfactory cues to maintain boundaries between lodges, which may range from about 1–20 per km of stream. These cues are created as mounds of mud and vegetation that may be small "mud pies" to those approaching 1 m in height; usually there are 2–8 piles per colony, which are located along water edges. This makes sense because beaver confine their movement to water courses, e.g., stream, lake, or its edge. By depositing scent from castor glands to signal occupancy of a lodge; by placing these along water edges, communicate to others in area, as well as to floaters.

A controversial topic of discussion among biologists is whether deer (family Cervidae) are territorial (Ozoga et al. 1982)? Territoriality in large herbivores is rare; it is documented only in roe deer (*Capreolus capreolus*) of Europe. Many biologists concur that female white-tailed deer defend an area about one-third the size of a home range (2 km; Lindemayer and Nix 1993) around a fawn against other female deer and males during the "hiding phase" of the fawn (about 4 weeks subsequent of birth) (Ozoga et al. 1982). This area around a fawn is called a fawning area, and its defense may help ensure adequate resources or to strengthen the bond between the mother and its young?

10.3 Problems in Defining a Territory and Ontogeny of Territoriality

Each animal requires is a set of resources unique to that species, so we would expect that there is diversity in types of territories. Some territories are for mating, some for nesting, while others serve multiple purposes (Brown 1973). A territory that serves solely for acquiring mates often is termed a lek, which has been discussed earlier. Defense of a nest site is common in colonial birds.

The ontogeny (e.g., origin and development) of territorial behavior in wildlife is unclear, but it makes little sense for young animals that do not breed to waste time and energy in being territorial (Barash 1973). My experience with eastern chipmunks is that territorial behavior in juveniles is nonexistent (Yahner 1978b). When juvenile chipmunks initially emerge from a natal burrow system, there likely is no way that they know the boundaries of a territory. In fact, rather than exhibiting aggression to siblings, juvenile chipmunks engage in play behavior. In contrast, adult chipmunks are strongly territorial in optimal habitat. Perhaps territorial defense only becomes when a burrow system with a winter, defendable larder hoard of food are established (Yahner 1978). Prairie dogs are born into a social organization that tolerates their presence for first week or two, then residents become more agonistic toward them, and the young prairie dogs soon learn residents from nonresidents.

10.4 Is Territoriality Genetically Programmed and Static?

Territoriality has been best studied in birds (Gill 1990). It was once thought that a territory was genetically determined and static. But then it was shown that great tits (*Parus major*) of Europe forego defense of winter territories on the coldest days as a means of conserving energy.

Optimal habitat for the North American red squirrel (*Tamiasciurus hudsonicus*) is coniferous forests. These squirrels typically are solitary and defend a territory to protect a stored source of cone seeds, known as a midden. In midwestern shelterbelts and perhaps in other areas of their range (other than boreal forests) where food is abundant, diverse, and not defendable, territoriality breaks down (Yahner 1980c). In shelterbelts, red squirrels are very tolerant of one another and do not have territories.

In a similar way, on the Galapagos Islands, the only ocean-going lizard, the marine iguana, feeds on marine kelp, which is superabundant, has no advantage to having feeding territories (Barash 1973). On one island, however, nests are limiting, and females defend nest sites.

10.5 How Do Animals Know Neighbors?

There is some evidence that territory holders learn to recognize neighbors by song in the case of birds and scent in the case of mammals (Davies 1978). In many passerine birds, knowing neighbors by song probably is adaptive. If neighbors learn the boundaries of their territories and those of neighbors, then they conceivably pose less of a threat to neighboring territory holders compared to strangers. Also, once boundaries are established, probably can maintain territory with limited effort against known neighbors (as mentioned earlier), thereby reducing a major expenditure of time and energy involved with territoriality (Barash 1973).

10.6 How Large Is a Territory?

As with home ranges, size of territories tends to increase directly with body size of in mammals. This suggests that bigger mammals are more mobile. Food availability also may dictate territory size. For example, Pomarine jaegers (*Stercorarius pomarinus*), defend a territory of 19 ha when lemmings (*Lemmus lemmus*) are abundant, whereas territories of jaegers increase to 45 ha when lemmings are scarce. Similarly, size of bird territories declines with increased density of flowers or presence of feeders.

Size of territory also may be a function of population density, e.g., territories may be packed together and, hence, be smaller as density increases (Brown 1973). In song sparrows (*Melospiza melodia*), territories become packed and smaller; as new birds arrive, areas left over in a given habitat may be too small for a territory (Davies 1978). Thus, territories in this bird may be established on a first-come basis. Biologists need to know size and location of territories in relation to other species as well. A case in point is the Serengeti Cheetah project (over 30 years) that has shown cheetahs are unique among large predators. Cheetah survival of cubs is low, being on about 75% due to predation by African lion and spotted hyenas. So, cheetahs avoid lions and hyenas and large prey herd, which attract these predators of cheetahs. Cheetahs are semisocial, with females either living alone or her cubs, whereas males coexist as groups of 2–3. Home ranges (or territories?) of males are only about 10% the size of those of females. Because of cheetah's avoidance of African lions and spotted hyenas, biologists have difficulty in finding cheetahs and, hence, understanding their behavior.

Size of a territory or even the existence of a territory may be a function of the presence of food. Laughing gulls (*Leucophaeus atricilla*) defend areas around picnickers and bathers on the beach; if gulls are not fed within 5–10 min, their territory disbands (Morse 1980). Glaucous gulls (*Larus hyperboreus*) defend eiders (*Somateria mollissima*) from other glaucous gulls; these gulls pirate food brought to the surface by eiders.

A question posed was whether animals defend super territories? In other words, do territorial individuals defend that are larger than necessary as a means of better ensuring survival and reproduction (Verner 1977)? However, there is no evidence for this spiteful behavior, based on studies of tree swallows (*Tachycineta bicolor*) (Davies 1978). Theoretically, the cost associated with defense of this large area (e.g., spiteful behavior) is prohibitive and this concept is not supported.

10.7 The Evolution of Territoriality

Territoriality is the most extreme and effective spacing mechanism (Brown 1973), which may explain why it has as a widespread phenomenon among wildlife (Morse 1980). Territories can be very diverse, making it difficult to find commonalities.

10.7 The Evolution of Territoriality

Some territories are for mating, nesting, or a combination of activities. Moreover, territories, as in many temperate songbirds, may not always occur in spring, but they may be present during winter or even year-round. For instance, song sparrows defend a territory in breeding season but maintain a home range at other times, but red squirrels defend a territory during all seasons. Furthermore, territories of red squirrels can change spatially, being found in boreal forests of northern latitudes, but not in deciduous forests in some parts of the Midwest (Yahner 1980c).

Territorial possession is often a prerequisite for successful mating; thus, there may be strong selective pressures for the evolution of some means to defend a territory via aggression, olfaction, or some other communication channel (Brown 1973, Morse 1980). A mating-territory or even a food-hypothesis makes some sense if either resource is both present in adequate quantity and economically defendable.

In order for territoriality to evolve, the benefits need to outweigh the costs of the defense of a space (Goodenough et al. 2001). There are obvious benefits and costs to aggressive behavior and territoriality. If an animal is aggressive, it could be exposed to risks of death and injury in battle and greater exposure to predation risks; this type of behavior time and energy to display and patrol boundaries perhaps could be devoted to more profitable behavior, such as that involved with finding food. Moreover, if territorial defense is risky, then why should an animal try to establish a territory, if failure ensures loss or not mating (Barash 1977)? However, a better approach to the question of whether territoriality is likely to evolve is not to look at the function of a territory, e.g., to acquire food or mates, but rather to examine differential fitness between those individuals with versus those individuals without territories.

Given the presumed costs associated with territoriality, is this adaptive for an animal (Barash 1977)? Using food as an example, if food is scarce or superabundant, an animal might waste time and energy defending this resource. Thus, territoriality might be favored when a resource is moderately abundant. The evolution of territory, using food as a model resource, was studied in red squirrel by Smith (1970). He reasoned that the aboveground larder hoard of red squirrels, termed a midden, is defended by solitary red squirrels rather than by more than one squirrel. A midden is economically defendable by a single squirrel because as food becomes depleted (cone seeds), the distance to cache food becomes much greater and costs a single squirrel 1.4-times more energy and time in territorial defense. In biological terms, this translates to a solitary squirrel spending 10% of its time caching food compared to 4% of its time caching if in a paired territory. Thus, for some species living in a group may be adaptive, but it could be detrimental to an individual of another species.

Red squirrels in eastern states are documented to cache hypogeous (belowground) or epigeous (aboveground) fungi [e.g., deer truffle (*Terfezia* spp.)], which are fungi that are buried 1–16 cm deep in soil, with the outer sporocarp ring, but they leave the center spore area. Truffles may be 45–95% of diet of red squirrels, depending on availability and availability of cone seeds. However, it is not documented whether this food resource is defended by red squirrels, as with pine cones in a centralized midden.

10.8 Interspecific Territoriality

If bald eagles (*Haliaeetus leucocephalus*) eat fish and song sparrows eat other items, why should there be territoriality between these two species (Barash 1977)? If both species fed on the same food resource, we might expect interspecific competition and, hence, interspecific territoriality to possibly exist between these species. At one time, similarity in appearance or voice, termed character convergence, was used to invoke interspecific territoriality. Interspecific territoriality may occur between species, but it appears to be much less common than intraspecific territoriality.

Studies of the great tit and the blue tit (*Cyanistes caeruleus*) of Europe suggest that predation may be related to the existence of interspecific territoriality because predators search for both birds at the same time (Goodenough et al. 2001). *Anolis* lizards' interspecific territorial defense does not seem to be a function of predation pressures but rather the size of lizards, regardless of species. Similar-sized lizards exploit the same-sized prey; thus, food overlap may sometimes dictate interspecific territorial defense. In three other examples, food may be a factor in the existence of interspecific territoriality. Mockingbirds may defend winter territories that have abundant berries (food) against other species that eat berries (Gill 1990). In pomacentrid fish (*Pomacentrus* spp.), over 90% of the intruders evicted from a territory by a given species are those where other species because use similar food. In limpets (*Eupomacentrus leucostictus*), defense of gardens of algae from both conspecifics and other limpet species that graze on these algae is common (Goodenough et al. 2001). This leads to the speculation that perhaps interspecific territoriality limits population densities (Yahner 2001). As discussed earlier, red fox minimize negative encounters with coyotes by placing their territories in areas devoid of coyote territories or along boundaries of adjacent coyote territories. As a consequence, interspecific interactions between red fox and coyote has probably reduced numbers of red fox; but red fox will continue to persist by occupying refugia habitats, e.g., urban areas, thereby fitting into the smaller areas, because a smaller red fox has a territory about one-third that of coyotes. Moreover, because body in red fox is smaller than in coyote, red fox have lower food-intake requirements of coyotes and higher fecundity (number of young per female). Existence of interspecific aggression (and territoriality) was examined between coexisting gray and red squirrel (Riege 1991). Aggression was not found between these two squirrels; instead, gray and red squirrels exhibit habitat segregation (deciduous vs. conifers, respectively). Thus, both species are not aggressive toward each other because separation is derived from habitat use and not via aggressive behavior.

Numerous studies, including those of mammals, birds, fish, dragonflies, etc., have shown that when territory owners are experimentally removed then their places are taken over by newcomers (floaters). The concept of a floater population developed in the 1950s when songbirds were removed from a forest to determine the effects of songbirds on insect pests (Barash 1977). The study was unsuccessful because as birds were removed, others took their place, which presumably were floaters and nonterritorial in the population.

In these removal experiments, it is important to determine where these floaters came from and did they reproduce without a territory. In great tits, replacements into woodland territories came from hedgerow territories (sink habitats) away from woodlands (source habitats); those from hedgerows had lower reproductive success than those from woodlands. In ruffed grouse (state bird of Pennsylvania), males display on a "drumming log" within its territory (Craven 1989). Thus, drumming birds are sedentary and presumably less susceptible to predation (stays within 100–150 m of log), with only males that drum have the potential to attract mates; thus, there is a floater nonbreeding male population. In some tropical birds, like the dusky antbird (*Cercomacra tyrannina*), experimentally created territorial vacancies, which are rapidly filled by neighboring birds. Whether these birds filling in vacated territories may be debated, but presumably the territory being filled by a vacancy offered better habitat is unclear. Sympatric white-bellied antbirds (*Myrmeciza longipes*) show mate fidelity and do not leave territories despite a vacancy created in a neighboring territory.

The presence of floaters suggests that prior to removal of territorial owners, e.g., residents, potential settlers presumably were prevented from occupying territories; this does not imply that territorial behavior has evolved to limit population densities (group selection of Wynne Edward, 1982), but that it had the consequence of limiting population density.

10.9 Introduction to Defense of Individual Distance

Territoriality may have evolved from the phenomenon of defense of individual distance. Defense of individual distance is the tendency for individuals to avoid proximity with each other so not provoke aggressive behavior (Brown 1973). Optimal spacing maximizes individual fitness; in striped skunks (*Mephitis mephitis*), for example, defense of individual distance breaks down while skunks use communal dens in winter. During winter (December–March), one male usually dens with several females (mean=6), as in an abandoned woodchuck burrow. Communal denning and a lack of the defense of individual distance enhance the energetics of skunks by increasing den temperature by 2–3°C for each animal, thereby saving stored energy in form of body fat during mild hyperthermia in winter. Skunks are solitary at other time of the year. Communal denning and a lack of the defense of individual also help to maximize reproduction in this solitary species because striped skunks breed immediately after emergence from a communal den, thereby ensuring successful reproduction. Close contact in winter is probably widespread in animals; close contact of individuals in severe weather is seen in some wrens (*Troglodytes troglodytes*) and in some swallows (Morse 1980). Thus, regardless of the species and its degree of sociality, all animals seem to maintain a characteristic individual distance; e.g., birds on a wire, but some species, termed species, like hippopotamus (*Hippopotamus amphibius*), essentially maintain no individual distance.

10.10 Other Differences in Defense of Individual Distance

Males typically have greater individual distances than females. In European rabbit (*Oryctolagus cuniculus*), subordinate males stay at least 1 m away from dominant males, but females and subordinate only stay about a body length (30 cm) apart from each other (Morse 1980). Experimentally, defense of individual distance has not studied often in natural populations; but in laboratory populations of chaffinch (*Fringilla coelebs*) showed that aggression among individuals declined with greater individual distance, individual variation in distance differed between adult females, but defense of individual distance was greater in adult males than adult females.

Adult male eastern chipmunks defend individual distance even when they are territorial at other times of the year (Yahner 1978b). In eastern chipmunks, defense of individual distance occurs at neutral sites and the resource is either mast [principally, mast of oak (*Quercus* spp.)] in autumn, when chipmunks moved out of defended core areas to harvest mast or during the breeding season, when adult males encounter one another away from their burrow systems in search for estrous females. Thus, greater than one spacing mechanism may be operative in the same species, depending on the time of year and the resource.

Brown-headed nuthatches (*Sitta pusilla*) (Fig. 10.3) feed on pine seeds; when seeds are abundant, these nuthatches feed so close that they often touch each other; if food is scarce, they are intolerant of each other (Morse 1980). Mobile territories have been reported around mates or food, as in male Cassin's finches (*Carpodacus cassinii*) that defend females and areas around the females rather than a fixed space.

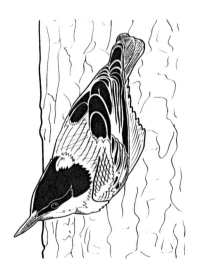

Fig. 10.3 The brown-headed nuthatch occurs in pine forests of the southeastern USA

10.10 Other Differences in Defense of Individual Distance

Perhaps mobile territories are found in species with male-biased sex ratios (number of adult males greatly exceed number of adult females). Mobile territories also may be found in adult male hamadryas baboons (*Papio hamadryas*), which are usually polygynous, with adult males guarding mobile adult females from other males. Thus, defense of an individual distance may even be considered as a "smaller" version of a mobile territory.

Chapter 11
Dominance Hierarchies

11.1 Introduction to Dominance Hierarchies

Territoriality and defense of individual distances are two spacing mechanisms typically involved with relationships between social units, whereas social dominance, as in hierarchies, is within social units (Barash 1977; Morse 1980). As with other spacing mechanisms, dominance typically establishes priority access to resources. Dominance usually is attained by animals that are most successful in aggression (e.g., via successfully fighting, chasing, or threatening a conspecific).

How does dominance differ from territoriality? Territoriality is a form of dominance, which is different from territoriality by the clear absence of a reference point in space, e.g., home or nest site. Dominance was developed as a concept developed by studies of domestic chickens (*Gallus gallus*) (Fig. 11.1) in 1922 by a German, Schjelderup-Ebbe (Allee 1938). He found that adult female chickens in flocks have a highly structured unilateral dominance hierarchy. This hierarchy is maintained via pecking (aggressive behavior) and has been termed a pecking order. Often the terms dominance hierarchy and pecking order are used interchangeably, but a dominance hierarchy if more general and means that an individual is typically dominant over another (e.g., statistically, the dominant animal more often wins than loses). As mentioned, a pecking order is always unilateral, with a winner always dominating a subordinate animal. A hierarchy tends to be more linear in smaller flocks or groups. Often, rather than being linear, a hierarchy is complex relationships, with triangles or squares, etc., forming among individuals in flocks or groups. A dominance hierarchy may be reversed at times, especially when initially it is established; usually, those animals in reproductive condition are dominant over those that are not in nonreproductive condition; this suggests that dominance is endocrine-based.

Dominance hierarchies are reported in mammals, birds, reptiles, amphibians, fish, arthropods, and mollusks, so this spacing mechanism is widespread (Brown 1977). As humans, we typically characterize a dominance hierarchy based on visual observations, but what about those that are established by animals using other means that we cannot detect, e.g., those established via olfaction?

Fig. 11.1 The chicken was domesticated, probably over 10,000 years ago in Asia. It is now a primary source of food and eggs for humans

Fig. 11.2 The eastern chipmunk is a small squirrel of eastern North America. It is solitary (except during its mating season, a female with young), diurnal, and lives in an extensive burrow system

In the wild, hierarchies may not be linear but rather nonlinear or relative, as in male eastern chipmunks (Fig. 11.2) during mating chases (Yahner 1978b). In relative hierarchies, individual males probably do not recognize each other (visually or olfactory) because time in association may be short or ephemeral. For instance, dominance hierarchies that form among adult male eastern chipmunks during mating chases of estrous females may include males that come from considerable distances and male composition of chases may change from chase to chase.

11.2 Advantages of Hierarchies

Hierarchies are not merely an alternative to territories, but animals in XXX to food has been shown in many animals, e.g., rhesus monkeys, groups or loose colonies confer advantages by defending feeding areas (coyote packs and carrion) or enabling better access to food (Barash 1977; Morse 1980). It has been found in juncos (*Junco* spp.) and field sparrows (*Spizella pusilla*) that high-ranking birds survive longer than lower-ranking birds; perhaps differential survivorship is because subordinate individuals have less access to food and are more likely to emigrate.

In fish and insects, for instance, a dominant individual may not appropriate access the food of a subordinate, but the dominant animal may have access to the best hunting places (Bertram 1978) or access to mates (e.g., alpha male in packs of gray wolves). In either case, a dominant animal spends less time and energy on aggressive interactions with other individuals of the same species. The relationship to dominance and increased access to females of woodland caribou, bison, mountain goats (*Oreamnos americanus*), and red deer (*Cervus elaphus*) is found in hierarchies. The dominant male of a harem does most of the mating, whereas subordinate males mated only before and after the main breeding season. In African hunting dogs, for example, 76% of the litters are produced by dominant males; also, 82% of the dominant females had litters each year. In contrast, litters were found only in 6–17% of the subordinate females. In gray wolves, only the alpha male and female tend to breed. How an alpha female wolf inhibits other females from coming into estrus or not breed is unknown (Bertram 1978). In contrast to gray wolves, all female African lions in a pride successfully mate, regardless of rank. But there are other examples where a subordinate male have success in mating, as with bighorn sheep in the Rocky Mountains; here, a subordinate male will challenge a dominant male in a head-butting contest for a few seconds of access to a female, which is enough to inseminate a female. As many as 42% of these subordinate males perhaps are successful in mating.

There is some evidence that animals living in a group may exhibit posttraumatic stress disorder (PTSD), as in the African elephant. Posttraumatic stress disorder is an anxiety disorder that can develop after exposure to a terrifying event or ordeal. Culling of dominant males and their transport to another reserve presumably was a terrifying event to subordinate males in the hierarchy. This leads to subordinate males initiating a killing spree, with the spree stopping only when older males were shipped in from Kruger, thereby establishing a new male dominance hierarchy.

11.3 Why Be Subordinate and Stay in a Group?

Perhaps there are four reasons why subordinate animals exist in a group (Goodenough et al. 2001). First by staying in a group, a subordinate may minimize competition with dominant animal. Second, leaving an established group and going to a strange area may be dangerous to a subordinate individual. Third, by staying in a group and helping raise young of a sibling may increase reproductive fitness of a subordinate animal. This may be the case in nonbreeding gray wolves or African hunting dogs and mongooses. In these canids, females often allow related offspring to suckle their young. Fourth, things might get better over time; what if the dominant animal dies, as in the alpha male of a wolf pack?

11.4 How Is Dominance Measured?

Dominance of animals over a subordinate(s) occurs in all spacing mechanisms (territoriality, defense of individual distance, or dominance hierarchies). There are at least two methods to measure dominance (Brown 1973; Morse 1980). One method is the use of paired encounters, whereby dominance is assessed in the laboratory using staged encounters in the lab. But a question often posed is whether theses paired encounters occur or can be extrapolate to wild conditions. A second, and perhaps more realistic, way of measuring dominance is to observe animal in either lab or field and watch the occurrences of aggression as they happen naturally. Clearly, a disadvantage of this method is that it can take considerable time to observe rare occurrences of aggression (Yahner 1978a).

Two issues that surface with either measure of dominance is what do you measure (Brown 1973; Morse 1980). For instance, with bighorn sheep it may be butting of heads or, with chickens, it may be occurrence of intraspecific pecking. In some instances, a subordinate may be supplanted by a dominant with no obvious mode of aggression. A second issue in that the observer of the aggression needs to precisely state how dominance is measured and in what context; this apparently is not always done. For example, in studies of primates, dominance may vary with context. For instance, in large enclosures, found that anubis (or olive) baboons (*Papio anubis*) mate frequently with females but are excluded by dominant males in mating with ovulating females. In the wild, individual yellow baboons (*Papio cynocephalus*) shifted dominance so often that advantages in mating are difficult to determine; it seems that female choice complicates the formation of hierarchies and the subsequent effects on reproductive success. In numerous studies, e.g. sage grouse on leks, elephant seals, male dragonflies (order Odonata), there is close correlation with fighting or displaying ability and mating success.

Regardless of whether we are examining the occurrence of aggression, or any behavior for that matter, it is difficult to extrapolate from lab to wild, from study to study, or from species to species. For example, what happens with mate dominance when other resources, e.g., food, are superabundant versus scarce.

11.5 Maintenance and Establishment of Hierarchies

Often aggressive interactions are very rare in nature, with eastern chipmunks spending <2% of their time budget in aggression (Yahner 1978b). Because during mating chases of eastern chipmunks, aggression is prominent and because individual males

11.5 Maintenance and Establishment of Hierarchies

probably do not know each other, the dominance hierarchy established among males is relative. In an absolute dominance hierarchy, we probably expect high levels of aggression at onset, when hierarchy is being established, but it would be adaptive for aggressive interactions to decline over time once the hierarchy is established and individuals comprising the hierarchy begin to know one another. Thus, cues are used to assess dominance probably include individual recognition, as with olfaction in packs of gray wolves (Morse 1980). Alternatively, cues to assess probably are those that accurately depict dominance, e.g., size of body and use of displays. Cues used to assess dominance ideally should be those that confer advantage in fights or aggressive encounter (Dawkins and Krebs 1978).

Dominant animals tend to be larger than subordinate animals, there gaining access to resources (Morse 1980). Size of body is typically indicative of dominance, even in birds, e.g., dark-eyed junco. Larger juncos are usually dominant over other individuals; birds that do not flock tend to be lower in the hierarchy. In the tropics, many species of antbirds (family Thamnophilidae) at least 50 species of birds follow ant raiding parties; the largest birds control the central area of the ant swarm and obtain up to 50% of their food from arthropods that are flushed by these ants (Gill 1990). In other species, e.g., hermit crabs (superfamily Paguroidea), a smaller individual leaves immediately when encountering a larger individual. Thus, size may be indicative of strength, as head butts of African buffalo (*Syncerus caffer*) (Dawkins and Krebs 1978).

Auditory cues may indicate the size of an individual and, hence, dominance status of a larger animal. A larger animals often make lower-frequency threat signals; so, frequency of a vocalization can be an indicator of body size and, hence, strength and dominance (Dawkins and Krebs 1978).

Factors, other than size of body, may be used to indicate dominance. In all species, adults tend to be larger and dominant over juveniles (Goodenough et al. 2001). In squirrel monkeys (*Saimiri*), alpha (dominant) males had the highest testosterone levels. Time spent, or tenure, in a group, may influence dominance status. Attaining dominance may take time, so new members of a group, e.g., pack members of gray wolves, need some time to work their way to the top of the dominance ladder. However, in African lions, two or more males may have equal dominance status, so mating opportunities are on the basis of first come, first serve (Bertram 1978). Dominance may be lacking, in this case, because fighting might be too costly, a companion male is beneficial to ward off infanticide male lions, a given mating may not ensure pregnancy, or these males may be closely related. In rhesus monkeys, or macaques (*Macaca mulatta*), rank of juveniles typically is determined by rank of the mother. In Canada geese, rank of a male goose is determined by rank of his entire family when it merges as a larger group. Furthermore, if two flocks of dark-eyed juncos come together, members of the flock with the highest ranking individual all become dominant over the second flock (Goodenough et al. 2001).

In Steller's jay (*Cyanocitta stelleri*), being dominant over another individual is inversely correlated with distance from a nest site (Goodenough et al. 2001). Similarly, in eastern chipmunks, solitary adult resident chipmunks are dominant over conspecifics when the resident is near (<15 m) from burrow site (Yahner 1978b). Previous encounters also may dictate dominance. Because it is unlikely that dominance is established repeatedly among individuals, particularly in absolute hierarchies, prior encounters may influence dominance (Goodenough et al. 2001). This has been shown in copperheads (*Agkistrodon contortrix*) in the laboratory.

In many raptor species, e.g., great horned owls (*Bubo virginianus*), eggs in the same clutch may hatch at different times, and incubation begins with the laying of the first egg. Thus, first young hatched sooner and develop quicker, and young that hatch later may starve if food shortages occur, with dominance varying within a given family group (Gill 1990). As we all know, artificial selection can develop certain breed or subspecies to enhance food production, as in strains of chickens or dogs for fighting.

11.6 Is a Dominance Hierarchy Better than Other Mechanisms?

Three factors might cause a shift from territoriality to hierarchies, at least in rodents (order Rodentia). These factors are an inability to disperse, e.g., leaving its natal site, an absence of escape cover, or a high population density. Each of these factors may be related to crowding or perhaps with inability to defend against predators (Barash 1977).

This suggests that hierarchies are more efficient when resources are sparse, because animals would not have to defend a territory; rather, if in a hierarchy, an animal can spend more time and energy finding food. Alternatively, however, competition among members of a group may cause some animals in a hierarchy to not gain access to resources. If in a hierarchy, when conditions deteriorate too far, as when food becomes scarce, an animal in a hierarchy will need to disperse or starve. Thus, a hierarchy may ensure that at least some individuals will obtain an adequate amount of the resource and survive.

Is aggression inappropriate? (Barash 1977); apparently not, if it is adaptive by enabling animals to compete for and exploit resources. Individuals can compete for resources in either of two ways: via scramble or contest competition. Scrambles occur when each individual attempts to accumulate or use as much of the resource as possible, without regard to social interactions. In contest competition, aggression is involved and in many forms, ranging from displays to actual fighting. Thus, it seems that aggression is a part occurs in most living things. But is aggression inappropriate or undesirable in humans? I do not think that I can answer this. However, there seemingly are problems with human aggression in that technology has given humans access to some very lethal weaponry, aggression is that can be used often at a considerable distances, aggression often used by entire nations, aggression is often used for some very insidious reasons that have no relation to maximizing fitness, and aggression often taken beyond the needs defined by animals (humans) to gain access to resources (e.g., when will it stop?). Moreover, humans may exhibit behaviors that verge on aggression, e.g., using unkind words to friends and family and antagonism toward a stranger.

Chapter 12
Communication

12.1 Introduction to Communication

Everything that an animal does, from assessing location or quantity of resources to escaping from predators requires communication. Therefore, an animal needs sensory perception of a resource or some means of interacting with another animal. Communication in animals is the most vital component of the behavior, and it mediates everything that an animal perceives or does in its environment.

From a human perspective, there has been of considerable interest when humans were first capable of speech and does communication in humans when it first evolved simply serve for communication or was it used for or transmission of thoughts? Hence, did speech as we know it today begin with *Homo neanderthalensis* or later with *Homo habilis* (Fig. 12.1)? In a truest sense, communication is the interaction between two or more individuals (Brown 1973); but, I would contend that an animal certainly uses its communication systems to find its way in the environment. As an example, a small mammal might orient itself in its home range by its perception of the location of logs or other features on the forest floor. This perception of environmental features is a means of communication.

A problem in studying behavior is that there may not be an observable response to a stimulus in the environment. For instance, when I examined auditory communication via playbacks of barks in Chinese barking deer, some deer did not seem to respond to this bark outwardly (Yahner 1980b). But I never really knew if they detected a bark of a conspecific, based on some physiological response, e.g., increased heart beat. So, animals may respond to signals from others, but as a human investigator, we may not know how to measure a response or even know that it exists, in part because we are visually oriented animals. We must assume, however, that for a signal to evolve, it must have some benefit to the sender. As humans, we might consider visual, audition, olfaction, tactile, and electrical as the major types of communication. These systems are studied most extensively and can be perceived by humans.

Fig. 12.1 *Homo habilis*, which means "handy man," is closely related to modern humans, but is least similar to that species, with short (less than half the size) and disproportionately long arms

12.2 Introduction to Visual Communication

Most animals have functional eyes, with some exceptions; e.g., some cave-dwelling animals. However, eyes of vertebrates typically are different from those of some mollusks and arthropods (Pough et al. 2002). Obviously, vision seems to be well-developed in birds, with rapid flight and diurnal activity; in birds (class Aves), eyes are large and the brain is displaced dorsally and caudally.

Birds have very good visual acuity; e.g., birds can form sharp images (Pough et al. 2002). This could be valuable to a flying bird in locating a branch to perch. Similarly, lizards (class Squamata) have very good visual acuity. In contrast, mammals evolved as nocturnal animals, so they have very good visual sensitivity, e.g., can form images in dim light. Retinas in eyes of mammals tend to have many rods; many mammals also have a well-developed tapetum lucidum, which is a reflective layer behind the retina that reflects light through the retina to enhance light detection at night (Vaughan et al. 2000). The tapetum lucidum causes eyeshine in many mammals, but this structure is lost in diurnal primates. Sharks (class Chondrichthyes) also need well-developed vision at low light intensities (Pough et al. 2002). This taxon has many rods, as well as a tapetum lucidum for detection of prey at night or at great depths. Humans have often looked at animals as a means of enhancing vision (night scopes, field glasses) or making vision more cryptic (camouflage clothes or weaponry).

A visual signal can be very ephemeral, e.g., the snarl of a dog or ear postures (Ewer 1973); thus, a visual signal may last only as long as the signal is being sent. In contrast, a visual signal can be long-lasting, as in claw marks left on a tree by a black bear.

12.3 Roles of Visual Communication

Because humans are visually oriented, I will start with this channel of communication. One role attributed to visual communication was to enhance group cohesion (Eisenberg 1966). It was thought that wing pattern of birds, e.g., geese (family Anatidae), was for group cohesion. Schooling in a tropical fish, e.g., golden pristella (*Pristella riddlei*), may be facilitated by marks on dorsal fins (Goodenough, McGuire, and Wallace 2001). How do marks on dorsal fins relate to the more familiar occurrence of tail flagging in white-tailed deer (Fig. 12.2)? Emphemeral visual signal, much like the snarl of a dog. Hence, a tail flash is an instantaneous signal. To best understand the role of tail flagging in deer, it might be worthwhile to hypothesize why tail flagging evolved. This visual signal may have evolved as an appeasement signal by subordinate individuals to dominant individuals (Smythe 1970). In some cases, the dominant animal may be a predator. Because tail flagging is given by both solitary deer and herd-forming deer, tail flagging is not believed to be used as an alarm signal.

Can a tail flag serve as a startle device, like the noise created by a ruffed grouse (*Bonasa umbellus*), as it flushed under your feet when you approach? A tail flag or a flush that momentarily startles a potential predator could be viewed as a startling device. A tail flag probably startles a potential predator, but why does a fleeing deer continue to tail flag well beyond the reach of a predator (after the tail flag served as a startling device)?

Fig. 12.2 The white-tailed deer has expanded its range and now occurs in all 48 states (except perhaps Utah) in the USA, Canada, and as far south as Peru. This deer has been introduced into New Zealand and some European countries

Tail flagging by white-tailed deer conceivably would enable a matriarch social group of deer to keep together (Smythe 1970). When flagging, a deer enhances its tail display by easy bounding compared to rapidly escaping from a predator. Thus, at least in deer, there seems to be some merit for this hypothesis.

If tail flagging in deer is used as an alarm or group cohesion signal, then why is this signal given toward a predator and not toward conspecifics (Smythe 1970)? Also, if an animal or a species is solitary, then why signal at all? Thus, if crypsis fails and a predator detects you, then make yourself conspicuous to the predator, particularly if you are out of attack distance. Support for this hypothesis is that a tail flag is given by both social and solitary individuals and species, a flag is given toward the attacker, and a flag at a safe distance. An advantage to this strategy is that a predator can be kept in sight at all times, so that the prey is not taken by surprise, especially in forested situations. This can be important because predator, if lost from sight, can sneak up on prey again, which would be a waste of time and energy to a predator.

Fawns less than 7 months old tail flag 50% more often than older deer. Hence, tail flagging may be a mechanism for fawns to remain with their mothers. Yet, tail flagging persists into adulthood because it may be a "risk-free" behavior.

A second role of visual communication may be for reproduction purposes. In Old World monkeys, like baboons (*Papio* spp.), have bare (hairless) rumps (called ischial callosities), which are multi-colored (red, blue, and purple) (Vaughan et al. 2000). These rumps are presented in precopulatory displays. Other roles of visual communication may be individual or species recognition, alarm, aggressive and defensive threats, camouflage, formation of pairs, or maintain pair and parent–young bonds. In the case of maintaining parent--young bonds, the gaping of young birds in the nest act as a signal for parents to feed the young, which has been well-studied (Gill 1990). Young birds with the proper and biggest gape marks receive most or all of the food from the parents. Cavity nesting species sometimes have brightly colored mouth markings to attract parental attention and to serve as targets for food delivery.

12.4 Spatial and Temporal Contexts of Visual Threat Displays

A territorial animal typically is dominant near a home site or other critical resource that is defendable (Brown 1973). Thus, visual threats tend to occur in a certain portion of the territory. From a temporal perspective, a territory owner typically precedes an attack with a threat; in other words, there is not an all-out war if an intruder comes onto a territory. This toned-down defense of a territory is demonstrated in bitterling fishes (*Rhodeus*) in which a male defends the mussel (class Bivalvia) while the female fish lay her eggs in the gill chamber of the mussel.

Also, from a temporal perspective, seasonal changes may occur in relation to the breeding season; songbirds are not territorial during nonbreeding season (typically), so threats and chases are confined to nesting season; in eastern chipmunks, threats and chases are highest during mating chases (Yahner 1978b). Threats probably help a territorial animal because hostile interactions can result in injury or death, and so

Fig. 12.3 The name mussel is used for members of several families of clams of bivalves from fresh and saltwater habitats, but most commonly used in reference to edible bivalves of saltwater habitats. One species, *Ligumia nasuta*, attracts fish using the display of a white spot on papillae that moves on the mantle of the female mussel

displays are often ritualized (Morse 1980). Signals that necessitate a threat display may vary with the degree of sociality of a species. For instance, on one end of the sociality spectrum, a solitary species may exhibit a threat display solely based on the presence of another animal. In contrast, a more social species may need not only the presence of a rival and also other information may be necessary to produce a threat display, e.g., large body size of a conspecific.

12.5 Types of Threat Displays and Decision Making by Animals

Bitterlings are known to have more than one threat display, and use of various threats differs with size of the opponent (Brown 1973). Three displays are used if an opponent is similar (jerking, turning beat, and fin-spreading). If an opponent is smaller, then only two displays are used by bitterlings (chasing and head-butting). In invertebrates and ectothermic vertebrates (fish, etc.), there usually are clear differences in body size between opponents; but in birds and mammals, opponents tend to be similar in body size.

An animal is constantly making decisions. For instance, is a threat warranted again in the presence of a rival or should the animal flee from the rival? At the boundary of a territory, there seems to be a balance between an attack and an escape tends to be about equal. Thus, at the boundary of a territory, the balance of an attack versus an escape may be tipped by small things, e.g., being slightly larger than the opponent. In addition, we might expect that the animal initiating the threat would be at an advantage in an attack or a chase.

Some displays in animals are intended for other species (Corey, Dowling, and Strayer 2006). For example, the life cycle of unionoid mussels include a larva (glochidium), with a glochida attaching to the gills or the fins of a host fish, typically a largemouth bass (*Micropterus salmoides*). Female mussels, e.g., in *Ligumia nasuta* (Fig. 12.3), attract fish by using the display of a white spot on papillae that moves on the mantle of the female. When a fish is attracted to this display of a white spot, the mussel releases a parasitic glochida.

12.6 Evolutionary Origin of Displays

There are at least six evolutionary origins of displays (Brown 1973). For instance, barbary dove (*Streptopelia risoria*) raise their feathers as temperatures become more and more cool. Thus, changes in feather positions can help regulate body temperatures. But in thermoregulation, no significant differences occur among body regions in temperatures. In mammals, get erection of pelage (fur) via piloerection (Vaughan et al. 2002). Many mammals erect fur when cold or when displaying, which makes the animal larger. Humans often get "goose bumps" when they are cold or scared. Perspiration and blushing may have been derived also in humans from vascular changes in thermoregulation.

Another evolutionary origin of threat displays may have come from respiration (Brown 1973). Air sacs of frigate birds (family Fregatidae) are brightly colored and used in courtship, and noses of proboscis monkeys (*Nasalis larvatus*) are important in facial expressions. Some species seem to give intentional locomotory movements that resemble an animal starting to escape. Some birds, as in ruffed grouse, raise or spread their tail or wings when alarmed. Saturniid moths (family Saturniidae) give rapid flight movements when a predator is nearby, which resembles original flight movements. Some butterflies have evolved large eyespots on their hind wings, which resemble the eyes of a vertebrate predator, e.g., snake (order Squamata). Some facial expressions that may be protective movements in preparation for biting, as in ear flattening, partially closing the eyes, and withdrawing corners of mouth, e.g., cats (family Felidae). Lowering of eyebrows in primates, e.g., gorilla (*Gorilla gorilla*), may be protective of the eyes, but now is ritualized as part of the stare used by dominant animals as a threat against subordinates. Certain displays, e.g., the swoop-and-soar in black-billed gull (*Chroicocephalus bulleri*), may be a re-directed attack one time. Out-of-context behavior, otherwise called displacement behaviors (Tinbergen 1940), are shown also by male mallard (*Anas platyrhynchos*) in the middle of courtship, with no apparent mating or dominance. These out-of-context or displacement behaviors tend to be incomplete and nonfunctional, are of greater intensity and rate than given original contexts, and tend to occur when two incompatible acts, such as advancing or retreating, are equally probable.

12.7 Evolution of Tusks in Walrus

Walrus (*Odobenus rosmarus*) are monotypic, meaning there is only one species in the pinniped family Odobenidae (Ronald and Gots 2003). Walrus once occurred along both coastlines of the arctic in North America, forming large amorphous herds; mating takes place during migration in the water; no harems formed on land, with the defendable resource being space on land. A conspicuous feature of males, in particular, is tusks, which are enlarged upper canines. In males, tusks may be 1 m in length and weigh 5 kg; which is longer and thicker in males than in females. Why did walrus evolve this conspicuous visual signal? Initially, tusks in both sexes were believed to be important in food acquisition. Walruses were believed to have evolved to rake mollusks (food resource) from bottom of ocean floor. However, this role of tusks is doubtful because walruses probably use hairs on snout, which number about 450 and are very sensitive to touch. Because walruses live in ice environments, a role of tusks has been propulsion while feeding on bottom or when moving across the ice, much like an ice pick. A fourth role attributed to tusks in walrus has been to break through ice to gain access to water. Because both sexes have tusks, it is suspected that a fifth and plausible role of this obvious external feature in females is to protect young walrus against two major predators, e.g., killer whale (*Orcinus orca*) and polar bear (*Ursus maritimus*). As mentioned, tusks in walrus are larger than those of females, and because, walrus are polygynous, tusks lively evolved for a very important sixth role, that being to enhance social status. Large body size and some external feature (here tusks) may be important in males gaining access to females during the breeding season.

12.8 Evolution of Antlers and Horns

Animals with hooves, e.g., ungulates, have evolved antlers and horns (Barrette 1977; Yahner 2001). Only a fraction of the total number of species of mammals have evolved structures on their head; in other words, antlers or horns are absent from mammals that live underground, e.g., subterranean or those that are aerial or aquatic. Yet structures on the heads of ungulates are prominent. Antlers are characteristic of male deer (family Cervidae) except found also in female caribou (*Rangifer tarandus*); the domestic version is the reindeer. Perhaps antlers are maintained in female caribou/reindeer as a calcium source. In deer, antlers are deciduous structures, which mean they are lost each year.

Five hypotheses have been given for the evolution of antlers in deer, with the first three receiving little support (Yahner 2001). First, antlers evolved to dissipate body heat because antlers are highly vascularized during the growth phase. But roe deer (*Capreolus capreolus*) of temperate regions of Europe have antlers in males that are still growing and are highly vascularized in winter. Antlers of the largest deer, the moose (*Alces alces*) of boreal climates, have antlers that harden during the heat of summer. Thus, antlers do not seem to help dissipate heat in deer. A second hypothesis for antlers in males to attract females also does not seem to be plausible. Females instead may be attracted to males with large body size. Larger, older males tend to be larger in deer than younger, inexperienced males.

Could antlers have evolved for defense against predators? If this is the case, then why do females lack antlers and why do antlers tend to be absent in winter and spring defense against predators tends to be greatest. Thus, there does not to be much support to this third hypothesis and the evolution of antlers in deer.

A fourth, and possibly plausible, hypothesis for the evolution of antlers in deer may be that it enables deer to assess dominance status of a rival. Rival males could use antler size to assess dominance and, hence strength, of an opponent, thereby reducing the need to fight and risk injury with a rival.

Possibly the most likely explanation of the evolution is its use in male–male combat. Small deer, e.g., muntjac (*Muntiacus*) use small antlers to knock an opponent off balance; once an opponent is knocked off balance, muntjac then strike with the opponent with enlarged canines. In larger deer, antlers can injure a rival male, so antlers serve to protect a deer from blows from the opponent, or antlers are used in pushing contests. It seems that antlers in deer are best developed in polygamous species where males defend a harem.

Compared to deer, members of the family Bovidae (e.g., sheep, goats, etc.) have structures (horns) growing permanently from the heads of only about 50% of the females. If females have horns, they tend to be found in females of species occupying open country, and horns of females tend to be straighter, thinner, and smaller than those of males.

Horns seem to have evolved for sex-specific reasons. In females, because horns are permanent structures and are thin and dagger-like, thereby serving as specialized stabbing weapons against potential predators. In males, horns are larger and often are spiraled or have some type of curvature. Horns of mountain sheep (*Ovis canadensis*) are large and curved, presumably to deflect and to absorb considerable force from the blow of an opponent during ritualized fights over females. Thus, rather that acting specifically for predator defense, horns of mountain sheep and many other bovids have evolved instead for intrasexual combat in between male rivals. An exception this dichotomy in horn structure and evolution of these structures likely exists in mountain goat (*Oreamnos americanus*). In this species, both sexes have dagger-like horns, possibly for predator defense; also, fighting is not ritualized between rival males.

12.9 Artificial Night Lighting

Too much of any type of pollution will have some effect on some type of organism. Light pollution probably fits into this category (Rich and Longcore 2002). We might define light pollution as the degradation of human views of the night sky, which is termed more appropriately astronomical night pollution. Wildlife biologists need to be concerned if artificial night pollution disrupts ecosystems or wildlife species of these ecosystems, which then is best ecological light pollution. Historically, humans often have used night lights, in part, probably using campfires to keep away predators. Sources of ecological light pollution today include night glow from natural sources, e.g., moon, stars, and glow from unnatural sources, lighted buildings, bridges, streetlights, lights on vehicles, and lights on fishing boats.

The extent of ecological light pollution is global, and it has been estimated that only 56% of Americans live in an area in complete darkness. Nearly 19% of the Earth is polluted by night sky brightness according to astronomical standards, with ecological light pollution typically is lowest in developing countries despite having the highest population densities; highest ecological light pollution occurs throughout most of Europe, North America, and some Asian countries, e.g., Japan.

12.10 Possible Effects of Ecological Light Pollution on Wildlife

All animals seem to be impacted by lights, as when lights disorient sea turtles along beaches or attract moths to lights (Rich and Longcore 2002). The familiar big brown bat (*Eptesicus fuscus*) of the northeastern USA, is the street light bat, which is attracted to these lights because they are adapted to feeding on hard-bodied insects attracted to streetlights (Whitaker 1972).

Most mammals are nocturnal (Vaughan et al. 2000), so ecological light pollution expected to have an effect on this taxa. Nocturnal mammals tend to react to increasing moonlight by reducing use of open areas, reducing time foraging, etc. (thereby, reduced predation risk); we found it interesting that most great horned owls tended to be most active at times during winter when it seemingly was easier for owls to see prey (e.g., 47% of our owl contacts when the moon was between the first quarter and full moon) (Morrell et al. 1991). A period of bright moon may be when most prey remain in secure places and, hence, less susceptible to predators. As possibly with great horned owls, ocelots (*Leopardus pardalis*) have lower successful prey encounters under bright light (Rich and Longcore 2002). Mountain lions tend to avoid urban night glow in southern California when exploring new habitats.

In white-footed mice (*Peromyscus leucopus*), full moonlight enabled animals to move greater distances (if predation risks were not a deterrent) to new uncolonized habitats (Rich and Longcore 2002). However, there is some evidence that small mammals (including white-footed mice) may reduce activity during bright moonlit nights (but see Barry and Francq 1982); thus, well-lighted roadways are likely to affect small mammal use of habitat near road edges. Moreover, small mammals (family Heteromyidae) decrease the amount of time foraging at night by 21% in response to a single camping lantern. In Florida, coastal lighting affected foraging by beach mice (*Peromyscus polionotus*) spend less time in habitats searching for seeds when bug lights were present versus absent (Bird, Branch, and Miller 2004). There is no support that increased lighting gives human drivers of vehicles more time to wildlife near a road, and thereby reduces deer--vehicular collisions. Each of us has experienced a glare phenomenon as someone shines a flashlight directly in your eyes at night; this results in being temporarily blinded because our eyes are cone-dominated, and rods are saturated with light. This also can affect animals with rod-dominated vision, e.g., deer, thus, causing "deer in the headlight syndrome."

Many birds also are attracted to light during migration at night. Perhaps light affects visual cues to the horizon and light, such as that from a lighthouse along the shoreline, is used as a cue for spatial orientation. Songbirds have been observed circling lighted platforms for hours or days, fall to ocean exhausted and emaciated, being unable to complete migration. Along British Columbia, mortality exceeded 6,000 migrating birds, which were noted at lighthouses. Some evidence suggests that fixed white lights are more deadly to migrating songbirds than colored lights or rotating white lights on lighthouses. A principal way of fishing for squid (order Teuthida) is by using artificial lighting on fishing vessels; squid have large eyes and are attracted to intense lights, perhaps because squid assume that food is going to be concentrated or perhaps the light make it seem that there are iridescent conspecifics. In one instance, over 6,000 crested auklets (*Aethia cristatella*) collided with a fishing vessel, nearly causing it to capsize. These nocturnal seabirds have a preponderance of rods in their eyes, making them susceptible to influences of artificial light.

The Florida coastline is major breeding area for two species of sea turtles: leatherback (*Dermochelys coriacea*) and green turtles (*Chelonia mydas*). The paradox is that the human population in this region has increased to 16 million in 1980, from only 1 million in 1920, which has affected choice of nest sites by females and hatchling orientation. Females typically use traditional nest sites, which are dark shorelines, unaffected by artificial night lighting. However, hatchlings on shorelines with artificial night lighting are either disoriented (crawl in circles) or misoriented (crawl toward light source). Disoriented hatchlings waste valuable time and yolk sources that should be used finding the ocean; in contrast, misoriented hatchlings are trapped in dune vegetation, are killed by cars, or are eaten by predators after sunrise. Florida officials have attempted to mitigate problems with turtle hatchlings on shore with artificial night lighting by installing lights with screens, using lower wattage in lights, or enforcing light curfews for outdoor or sporting activities.

In California, nocturnal snake populations are declining rapidly in areas with intensive human development (Harder 2004). The consensus is that desert snakes are most active under full night darkness; but this response may be age-specific. For instance, adult prairie rattlesnakes (*Crotalus viridis*) avoided lights because they feed largely on rodents, whereas juveniles feed on insects that are active mainly in day.

Juvenile common toads (*Bufo bufo*) presumably congregate under streetlights to capture insects attracted to lights (Baker 1990). There also is some limited evidence that Tungara frogs (*Physalaemus pustulosus*) are more selective of male mates in greater darkness, e.g., they make a "quick-like" choice of mates to avoid predation. Moreover, these frogs seem to hide nests better in brighter light. Some tree frogs, e.g., *Hyla squirella*, from Louisiana, stop or reduce choruses when disturbed by humans under full moonlight, or when affected by artificial lighting. Salamanders may be affected by artificial night lighting, but most studies of this taxon have dealt with lab populations (Wise and Buchanan 2002). Larval American toads (*Bufo americanus*), known as tadpoles spend nights in deeper water at night because deeper waters at night are warmer; but during the day, tadpoles move to shallower water where it is warmer. Thus, if artificial light were present, may lose natural light cues and, hence, their ability to thermoregulate by moving into colder waters lighted by artificial lights at times when these waters are cooler.

Fish tend to school when light levels increase, perhaps to avoid predation risks, but foraging at lower light levels as individuals is less efficient (Nightingale et al. 2002). Rainbow trout (*Oncorhynchus mykiss*), for example, seldom forage when there is a full moon or artificial light. Plankton tends to move to the surface when it is dark and descend during the day, so planktivorous fishes can increase foraging efficiency. Similarly, when the moon rises late in the night, what started as a night in complete darkness, enhances the foraging efficiency of these fish, which is called a lunar light trap. During the migration of sockeye salmon (*Onconchus nerka*) fry, artificial lights from shoreline buildings and bridges causes them to delay migration and swim to lighted shorelines, where water velocity is lower.

Many insects, e.g., moths (order Lepidoptera), have no apparent way to resist light (Frank 2002). Moths, for instance, fixate on the light by continually flying around it to exhaustion or movements across the landscape are interrupted, so light acts as a barrier. Distances of attraction to lights can exceed 100 m, depending on the species. Swarming species are particularly affected by artificial lighting. Light traps have been used to attract and subsequently identify insects.

Chapter 13
Olfactory Communication

13.1 Introduction to Olfactory Communication

We might be tempted to lump olfaction, which is smell, with taste into the chemical sense because these two senses are closely linked. For instance, have you ever tried to taste food when you have a severe cold, which blocks the sense of smell (Brown 1973; Pough et al. 2002)? Moreover, both smell and taste involve the detection of dissolved molecules by specialized receptors, but each has a very different embryonic origin. Smell is a somatic sensory system, with sensing at a distance (a skunk smell; Yahner 2001), and the sensation of the smell is sent to the forebrain. In contrast, taste is a visceral sensory system, e.g., this sense deals with the direct contact of sensations that are sent initially to the hindbrain. In this chapter, I deal with smell, or olfaction. This communication channel has been studied mainly in mammals and insects, but it has been studied somewhat in other taxonomic groups.

13.2 Olfaction in Various Wildlife Species

In sharks (superorder Selachimorpha), olfaction is the first sense used to detect prey, as when a shark follows the smell of blood and its gradient to prey. Sharks and salmon know to detect odors at concentrations of less than one part per billion. Most fishes (infraphylum Gnathostomata) generally have taste-bud organs in their mouth, around the head, and even in anterior fins. Manufacturers of fish lures (Fig. 13.1) have incorporated flavors and scents into soft bait to enhance their attractiveness to fish. Sharks and salmon (family Salmonidae) are capable of detecting odors at concentrations of less than one part per billion.

For a long time, a belief was that olfaction was poorly developed in birds, but may not be true of all birds (Pough et al. 2002). However, some birds have large olfactory bulbs in brains. Some birds, such as kiwi (*Apteryx* spp.), have nostrils at the tip of their bill, which apparently, are used to locate prey in soil. Turkey vultures,

Fig. 13.1 A fish lure is an object that often is tied to the end of a fishing line and is meant resemble and move like fish prey. Because fish have a good sense of smell, many lures smell like fish prey

which are scavengers, can smell carrion. Water birds, e.g., shearwaters (family Procellariidae), can smell food even in darkness. These birds also approach islands used for nesting by smelling the islands, which they approach from downwind. Furthermore, homing pigeons (*Columba livia*) partially use olfaction to navigate.

Some animal have a unique organ of olfaction, called the vomeronasal or the Jacobson's organ, which is located in the roof of the mouth. When snakes, for example, flick out their tongue, they transfer molecules from air to this organ (Brown 1973). Ungulates exhibit a behavior, termed flehmen, in which a male curls the upper lip and often holds its head high, enabling him to probably inhale molecules of chemicals in the urine of females; this allows the male to transfer these molecules to the vomeronasal organ and thereby assess the reproductive status of a female. Primates (order Primates) with relatively flat faces were thought to have lost the vomeronasal organ, but there may be a remnant of it in primates.

In mammals, olfaction typically well-developed, which reflect their nocturnal way-of-life (Brown 1973; Pough et al. 2002). Mammals typically have large olfactory lobes, but these lobes are small in primates and absent in whales (order Cetacea), where olfaction is reduced or completely absent.

Olfactory communication is similar to hormonal action because a relatively small amount of a given chemical is used to send a message (Brown 1973; Pough et al. 2002). Many animals perceive the world through olfaction, and olfactory chemicals that carry information between individuals called semiochemicals. Allomones are used among species, whereas pheromones are used within a species. Allomones often are often used as repellents against other species, with the best-known example being scent produced by a skunk. Repellents also are well-developed in insects, as with the southern walkingstick (*Anisomorpha buprestroides*), which ejects a defensive spray when disturbed. On the other hand, pheromones, as sex attractants, are known in insects, crustaceans, fishes, salamanders, and mammals. The effectiveness of sexual pheromones has been used in controlling Japanese beetles in our yards, gypsy moths (use artificial lure, disparlure), and Mediterranean fruit flies. Some studies have noted that males respond to pheromones of females of other species, which suggests that these species are capable of interbreeding. Then why don't they? In these cases, species are not sympatric (occupy different geographic ranges) or mate at different times of the night.

13.3 Olfaction as Alarm Signals

When a pike (*Esox* spp.) injures a minnow (family Caprimidae), the minnow often releases a chemical, known as Schreckstoff, from damaged epidermis (Goodenough et al. 2001). This olfactory signal keeps conspecifics away from the damaged minnow for hours or even days. Destroyed tissue serving as alarm signals has been shown in several aquatic gastropods, in earthworms, and in sea urchins. In social insects, release of alarm signals is not dependent on tissue destruction. For instance, in formicine ants (subfamily Formicinae), it is a minor threat to enemies that involves biting and stinging the enemy with the release of formic acid; the alarm substance, however, is a chemical, called undecane. In fire ants (*Solenopsis* spp.), a trail pheromone can be created, which is volatile within a couple of minutes. This pheromone leads colony members to a food source; but once food is gone, the trail pheromone no longer a serves as a chemical signal. Mule deer have an alarm signal that is emitted through an enlarged metatarsal gland on the hind leg; this gland is smaller (about 5 × shorter) in the congeneric white-tailed deer (41 mm in length); and is nonfunctional in this latter deer species (Müller-Schwarze 1971). Despite the small size of the metatarsal gland in white-tailed deer, it may play a role in individual recognition.

13.4 Buck Rubs

Many signals may function as both an olfactory and a visual signal. For example, male white-tailed deer produce buck rubs about 2 months before the rut (breeding season), which occurs in the much of North America from about September through January (Marchinton and Hirth 1984). Buck rubs usually are placed on small trees with smooth bark, causing a problem in deciduous shade tree nurseries (Nielsen et al. 1982). Hence, a buck rub serves as a visual signal; it also serves as an olfactory signal because a buck rub is rubbed vertically with the base of antlers and forehead gland (sudoiferous, which is a type of sweat gland; Vaughan et al. 2000), which is well-developed in males during the rut.

A rub presumably sends six signals to other deer: individual recognition, sex class, sexual receptivity, age, because frequency of rubbing and development of forehead glands is greatest in older bucks, presence, and possibly a buck rub may induce and synchronize mating with several females in an area; evidence for this sixth function comes from observations that when female detects a rub, she licks and smells it. This sixth function is adaptive to a widely dispersed species, e.g., white-tailed deer, because a male deer can court and mate with several females. Incidentally, buck rubs were once believed to be adaptive to male deer by removing velvet and giving combat practice. Buck rubs probably do this for males but it is not why they evolved.

13.5 Buck Scrapes

Another signal that has both a chemical and a visual role in white-tailed deer is the buck scrape. Unlike a buck rub, the scrape is produced by male white-tailed deer later in the breeding season in North America and after leaf fall, e.g., in November (Ozoga 1989). To produce a buck scrape, a reproductively active male paws the ground to produce a visual signal, which is about 0.5 m in diameter, on ground along a well-used trail. Thus, a buck scrape is a conspicuous visual signal, but it is combined with olfactory signals by the same male defecating and urinating on the scrape and then marking an overhanging branch (1–2 m aboveground) with the forehead gland.

A buck scrape likely intimidate rivals, attract females, and may synchronize mating. Synchronized mating has been noted in mice (*Mus* spp. and *Peromyscus* spp.), which is produced by the presence of smell (of male urine), termed the Whitten effect (Eisenberg and Kleiman 1972). Besides synchronizing mating, olfaction may advance sexual maturity. This has been shown in pigs (*Sus scrofa*) and mice. Females of these species reach puberty earlier, when housed with males or exposed to male urine in the laboratory. Another phenomenon also occurs in mice in the laboratory, termed the Bruce effect. In this phenomenon, recently inseminated females may not become pregnant when detecting the odor of a strange male; if such a phenomenon occurs in the wild, this blockage of implantation may act to reduce population densities when population densities of mice are high.

13.6 A Universal Trend

There is a universal trend in mammals, whereby dominant individuals mark more than subordinate individuals. In common raccoons (Fig. 13.2), for instance, anal rubbing occurs most frequently in dominant animals at communal defecation sites. In gray wolves, scent-marking is mainly done by dominant animals; lone wolves seldom mark, but instead keep low profile (Rothman and Mech 1979). Newly paired wolves, however, mark most often to establish a new territory. Continuous scent-marking in dominant wolves helps form and maintain pair bonds.

13.7 Cheek Rubbing

Many squirrels, which are solitary and territorial, show cheek rubbing, including woodchuck (*Marmota monax*). Woodchucks twist their heads on one side than drag the side of the face (cheek glands) along fences, woodpiles, shrubs, rocks, burrow mounds, typically within 6 m of a burrow. Nonresident woodchucks have been observed marking over the mark left previously by another woodchuck at its burrow, suggesting that marking is for individual recognition and signaling that a given burrow

13.9 Scent-Marking in Canids

Fig. 13.2 The raccoon is usually nocturnal (active at night) and it has two distinctive features, a facial mask and very dexterous forefeet. Raccoons typically are solitary

is occupied. Thus, an unanswered question is whether an olfactory signal is simply used for self-assurance, e.g., I am in the right place and I have been here before, or that it becomes a dominance signal to conspecifics. Cheek marking in pikas (order Lagomorpha), which are related to rabbits live in alpine areas; cheek glands in pikas are about 4 cm below the base of the ear on the cheek. Like woodchucks, pikas are solitary, and male pikas select territories surrounded by those of females.

In northern short-tailed shrews (*Blarina brevicauda*), flank glands are well developed during the breeding season. In these shrews, flank glands probably help males to attract mates. Shrews are fossorial mammals that may have a difficulty in finding mates otherwise.

13.8 Olfaction and Infanticide

We have dealt with the phenomenon of infanticide in lions and bears earlier (El-Jaddad et al., JM 69:811). In many of these species, olfactory recognition of a female mate may be sufficient to prevent males from killing young in monogamous or polygamous species. However, promiscuous species, such as deer mice (*Peromyscus maniculatus*), may require that males know both the female and the young by scent, with neonates carrying the odor of familiar female. Thus, if an unfamiliar individual present based on olfactory signals, either female or young, infanticide is triggered.

13.9 Scent-Marking in Canids

There seems to be two advantages to a single olfactory signal, like a scent-mark created by a canid – it is long and site-specific (Eisenberg 1966). A disadvantage of a single scent-mark is that it lacks directionality. So, if scent-marks are used to delineate the boundary of a territory, a single scent-mark conveys limited information about the boundary of a territory. Scent-marks by domestic dogs are familiar to all of us, as the dog leaves a chemical signal in its urine on a telephone pole or fire hydrant (Yahner 2001).

In canids, scent-marks probably originated as a self-assurance signal, e.g., "I've been here before" or "this is a familiar area" (Eisenberg 1966). Later, this self-assurance signal evolved into a "too whom it may concern signal," giving information on individual identity, sex, age, reproductive status, and perhaps dominance status. Because a single scent-mark does not provide directionality, a lone gray wolf that encounters a scent-mark deposited by an established pack could be attacked or even killed if it opted to move in the wrong direction and head into the center of the territory of the wolf pack (Peters and Mech 1975), implying the need for a gradient. How does a gradient work? Members of a wolf pack travel along logging roads, trails, or other established routes within their territory at about 8 km/h. While in route and depending on their location, wolves vary the rate of scent-marking, which in effect, vary the distances between marks. Scent-marks are more concentrated along the edge than in the interior of a territory, being only 110 m apart along the territorial edge – this is about a 1-km wide border; in the center, scent-marks are spaced about 180 m, resulting in an "olfactory bowl." In addition, when members of another pack encounter marks left by a neighboring pack in this 1-km border, they scent-mark over the marks left by the previous pack, which, in effect, conceivably doubles the number of marks on the edge compared to the interior of the territory.

The territory of a wolf pack may be 10–20 km wide; thus, this 1-km border or edge between neighboring packs is visited less frequently than other areas of a territory, which reduces interpack aggression. In turn, the lower concentration of scent-marks in the interior of a territory helps members orient themselves within the territory, particularly if they become separated from their pack.

13.10 Mud Piles in Beaver

Few animals are as social as the American beaver. A beaver colony often consists of an extended family unit, living in the same lodge or bank den. During winter, 3–4 generations may live in the same lodge: an adult pair, one or more 2.5-year-olds, yearlings of 1.5-years-old, and kits from the previous spring. Colony size may vary with latitude, with 5–6 in southerly latitudes of USA and 3–4 in more northerly latitudes of Canada. However, not all beaver live in colonies; 15% live singly and 24% may live as a pair. New colonies formed by dispersal of 2.5-year-olds that leave a natal colony.

Density of individual colonies is maintained via territorial behavior (Yahner 2001). Because beaver are nocturnal or crepuscular, boundaries of a given territory are best maintained by olfactory signals.

Beaver use "mud piles" as olfactory signals; each mud pile may be up to 1 m in height, and olfactory signals are placed on these piles. Usually there are 2–8 piles per colony, which are placed along water edge near boundary of territory. Beaver mark piles using chemicals produced by their castor (hence genus *Castor*) and anal glands. Castor glands are on underside of tail and produce a yellowish substance that turns brown when exposed to air and sunlight. Castor is mixed with urine to form castoreum and is expelled through the cloaca without the beaver even touching the pile. Castoreum acts to delineate the territory boundary and elicit a sense of "confidence" or assurance among colony members, and it signals that a given colony is occupied to floaters in the population.

Anal glands also used on piles (Yahner 2001); a drop of pungent, straw- to brown-colored substance is produced and rubbed on the pile or scent mound. It seems that the functions of anal glands are to identify individuals and to waterproof the pelage of a beaver. Because scent mounds aid in maintaining territorial rights, population densities and intraspecific strife are reduced before food can become a limiting factor.

13.11 Use of Scent Stations in Mammalian Research

Because mammalian carnivores rely extensively on olfactory communication, scent stations have been used for decades by the wildlife profession to determine presence and possible abundance of elusive carnivores (Conner et al. 1983). A scent station is a 1-m radius cleared area, typically sandy or consisting of soil and devoid of vegetation and leaves, and scent placed in the middle as an attractant to carnivores. Because mammals mark over scents of another, scent stations can be used for a variety of sympatric species. Scent stations were developed originally in the early 1950s for red and gray foxes, but they are now used to monitor coyotes, bobcats, wolves, and other species. A recent study attempted to determine if scent stations can be used for estimating seasonal or annual abundance and if scent station data reflect population estimates obtained by other techniques (e.g., time consuming trapping). Results showed that visitation rates per month varied from 1% for bobcat to 48% for gray foxes; visitation rates were reflective of population estimates for bobcats, raccoons, and gray foxes, but they were not for Virginia opossums. Thus, some protocols (e.g., scent stations) seem to fit for some species but not for other species.

In studies of nest predation on birds, a lingering question has been whether investigators leave scent and thereby increase nest predation (Donalty and Henske 1991). Carnivores have been suspected of preying on bird nests, e.g., raccoons in central Pennsylvania (Yahner and Morrell 1991), but birds, such as crows are also major predators (Yahner and Wright 1985). An interest in studying nest predation in birds is twofold. First, predation is the major cause of mortality on nesting success in birds, and second, several investigators have shown that construction of artificial nests can depict mortality of actual nests. Artificial nests have been used in experimental studies because they can be placed under any set of conditions, and finding actual nests is time and labor consuming.

To reduce mammalian carnivores associating human scent with nests, investigators should wear rubber boots and rubber gloves when checking nests or when handling content of nests. I suspect that a trail created by investigators when checking a nest act as a visual signal and reduces energy in foraging or imply walking by a predator that is more likely to attract a foraging mammalian predator than a scent.

In American robins in Colorado, investigators handled chicks and eggs to leave human scent on these young robins. Conversely, chick and eggs were not handled for comparison. No effect of human handling was found on nesting success in robins in a comparison of these two groups.

Chapter 14
Auditory Communication

14.1 Introduction to Auditory Communication

Compared to most communication systems, an auditory signal can be transmitted over long distances; but depending on the frequency of an auditory signal, it may be hard to localize for reasons that we will discuss. What then is a disadvantage of an auditory signal? First, it may be short-term, e.g., when the sound stops, the signal stops.

14.2 Barking in Forest Deer

As with other communication signals, functions of auditory signals can be divided into those with value to the individual (sender) or to the group in general. In a solitary species, why be vocal and call attention to yourself? Perhaps vocalizations in solitary animals simply are external expressions of internal anxiety? (Yahner 1980b). Using the barking deer, or Chinese muntjac (Fig. 14.1), as an example. In the wild and in semicaptivity, these animals bark at novel or inconspicuous objects in their home range. In the wild, it has been reported that large predators, such as tigers, will hunt elsewhere if detected by a prey (Schaller 1967). Thus, barking in Chinese muntjac sends a signal to a stalking or hidden cat that it has been spotted (Yahner 1980b). The result is that bark from a deer becomes a predator deterrent and is likely an internal expression of anxiety. Animals, especially visually focused ones, know their home ranges and its features very well, which has adaptive value. Barking is a better response to a predator than simply running away at a safe distance because a predator can then sneak up on it again. If the object is novel or inconspicuous and does not move, then the muntjac can simply ignore it and continue other activities, such as foraging.

Why can this bark not serve as an alarm signal to a conspecific (Yahner 1980b)? Muntjac are solitary deer, and young disperse early; so, as an observer, I cannot rule out that the signal is interpreted as an alarm signal prior to young dispersal, but

Fig. 14.1 Reeves' muntjac, or Chinese muntjac, is given its common name from John Reeves, who was an Assistant Inspector of Tea for the British East India Company in 1812, and because this species occurs in China

I would question otherwise to whom is the message being sent? I found no evidence that there is individual recognition of individual barks, based on use of playback of a muntjac bark. Furthermore, the muntjac is a forest deer, and its bark is relatively low in frequency, which enables a predator to easily localize its source; plus, low frequency calls are effective in penetrating the dense vegetation of a forest, which then supports the idea that barking in Chinese muntjac is actually a predator deterrent and is likely an internal expression of anxiety.

14.3 Barking and Domestication in Domestic Dogs

Mitochondrial DNA sequencing that has been done relatively recently has shown that domestic dogs originated from gray wolves about 135,000 years ago; once thought to have been domesticated 10,000–12,000 years ago (Vilà et al. 1997). Why was the gray wolf the best ancestor of domestic dogs? The answer probably has a strong behavioral basis, which is especially intriguing because gray wolves are genetically similar to other wild canids, with red wolves, coyotes, and four species of jackals, can interbreed and have 78 chromosomes (all are within the genus *Canis*).

Gray wolves are circumpolar and are in close association with human populations; thus, domestication may have occurred in different times and places in the world (Yahner 2001). Once domesticated, dogs rapidly spread throughout the world as companions to nomadic humans. But dogs perhaps become only morphologically different (visibly) from wolves only about 10,000–15,000 years ago as humans went from a nomadic to an agrarian way-of-life. Today, dogs are selected artificially by humans on the basis of specific traits, e.g., toys with small facial features (Chihuahuas, Pekingese), sporting companions (retrievers, spaniels), or guard dogs (rottweilers, German shepherds).

The gray wolf was not the only canid in close contact with humans in prehistoric times, although coyotes were presumably attempted to be domesticated by native Americans, jackals in Africa and Asia, and the extinct Falkland Island wolf, which probably was very tame canid and seen by Charles Darwin in his journeys to the islands.

A major behavioral aspect that gray wolves had was a hunting style that was similar to early human groups. Both groups hunted cooperatively, and wolves accompanied humans and assisted in finding and capturing prey with excellent senses of olfaction and audition. In addition, both early humans and wolves lived in small groups, consisting of related individuals from several generations; members of a group coexisted as dominance hierarchies, with humans providing subordinate wolves with food and other necessities. Hence, in this combined social groups, domesticated wolves imprinted on humans as their group leaders. Possibly adequate provision of food to dogs by humans is why dogs have 2 l per year compared to 1 l per year in wild gray wolves. Domestication of gray wolves may be advantageous to early humans in two other ways. First, as with some aborigines (native peoples of Australia) and dingoes (wild dog of Australia), domesticated wolves and human likely slept together; sleeping together communally provides warmth and protection. Also, wolves howl and bark, so humans in small, cohesive groups probably selected individual "watchdogs" that barked frequently in response to possible danger.

14.4 Vocalizations in Wildlife

To begin with why do gray wolves and coyotes howl and what information is conveyed? Howls can be heard for about 2 km in the wild, and they serve both intrapack and interpack functions. Even though all members of a wolf pack are capable of howling, only the alpha (dominant) male howls. Howling by the dominant male (usually) serves an assembly call or intrapack function to reunite pack members. In contrast, a subordinate animal, when alone away from a pack, will not howl; instead, it keeps a low profile so not to attract a strange pack and be killed. Interpack howling, which includes all or most pack members, is given in two general contexts when approaching an area used jointly by a neighboring pack, as in the 1-km outer border of the olfactory bowl, thereby advertising the presence of a pack to a rival pack, if the rival pack is in the immediate area, or when the pack returns to a portion of its territory that has not been visited in some time, suggesting that olfactory scent-marks had lost their effectiveness in establishing territorial boundaries.

Coyotes produce two types of vocalizations, not unlike the gray wolf, a group howl and a yip-howl (Lehner 1982). The group howl is similar to the howls given by the alpha male wolf to reunite pack members. The yip-howl is given by all pack members to signal to a nearby pack that a given territory is occupied; hence, the yip-howl of coyotes acts like the group howl of wolves.

Most likely vocalizations act as alarm call in some species, such as in social animals, where another member of the species group is near to hear the alarm call (Yahner 1980b). Social deer of Asia, e.g., Axis deer (*Axis axis*), respond to alarm calls given by other Axis deer and to those given by a coexisting species, e.g., sambar (*Cervus unicolor*).

A vocalization may act as a "to whom it may concern" signal (Eisenberg 1966). For instance, a given signal may convey the age of the animal. A younger animal typically has a vocalization that is higher pitched usually in younger animals; for example, fawns of white-tailed deer emit a high-pitched bleating (Richardson et al. 1983). Sometimes, the frequency of a call is characteristic of the sex of an animal. In great horned owl (*Bubo virginianus*), the "hoot" is lower in the larger female than in the smaller male (Morrell et al. 1991).

Some vocalizations may act as dominance signals. The dominant male of a white-tailed herd gives a series of aggressive snorts to signal dominance over subordinate males (Richardson et al. 1983). Most calls are species-specific and may be the only way investigators can tell the identity of a species, e.g., *Empidonax* flycatchers (Gill 2000).

Another "to whom it may concern" signal may be that of species identity. It is adaptive for individuals to know the identity of neighbors and, hence, not respond as rapidly and decisively to a call of a neighbor compared to an intruding nonresident (Yahner 1980b; Gill 1990).

In deer and numerous other animals, a certain vocalization may serve as a reproductive signal. For instance, a courting male muntjac typically gives a soft buzzing call, which presumably allows the male close proximity to an estrous female (Yahner 1979). Male eastern chipmunks also give a squealing sound when chasing estrous females during mating chases (Yahner 1987b). The buzz of the male muntjac and the squeal of the male eastern chipmunk are not given in any other context, to my knowledge.

14.5 Bird Vocalizations

Birds give a variety of vocalizations, e.g., alarm calls to signal danger or to initiate escape behavior (Gill 2000). Contact calls, which may be similar to tail flagging in deer, enable birds in a foraging flock to keep track of one another in dense vegetation. As with muntjac or eastern chipmunk, precopulatory trills and postcopulatory grunts are heard only during mating displays.

Birds also give complex territorial songs (Gill 2000). These songs are usually by conspecifics for at least 50 m or more. These vocalizations give information "to whom it may concern," e.g., individual identity, species identity, and reproductive motivation. But songs also signal to potential rivals that a territory is occupied, and they probably attract females for courtship and pair formation. Once a male is mated, it sings frequently once a mate is acquired; e.g., rates of singing in unmated ovenbirds and Kentucky warblers were 3.5 and 5.4 times higher, respectively, than mated birds (Gibbs and Wenny 1993; Yahner 2000). If a female is experimentally removed from a pair, there may be a resurgence in song rates by males. The chance of pairing may be aunction of the amount of forest and other features; thus, so songs might be frequent, if a male is not mated (Gibbs and Faaborg 1990). For example, in forested areas of central Pennsylvania, the chance a forest warbler, the ovenbird (*Seiurus aurocapillus*) (Fig. 14.2) is negatively correlated with % forest cover within 1 km (e.g., 50% mated if 95% forested, 65% mated in 80% forested) (Rodewald and Yahner 2000). In Maine, 47% of ovenbirds mated in smaller woodlots (<70 ha)

Fig. 14.2 The ovenbird is a small, ground-nesting bird of North America. This songbird migrates south as a winter strategy

compared to 67% in larger woodlots (>70 ha) (Yahner 2000). In American redstarts, males sing one song in repetitive fashion (repeat mode), but alternate between songs when mated later in season (serial mode) (Stacier et al. 2006). Although not studied, I suspect that this phenomenon is common in other species that use alternate songs.

Based on playback studies, bird probably seek out songs of individuals of the same species by approaching the location of the new song or begin singing in response to this recording (after Gill 1990). In central Pennsylvania, playbacks have been used to increase the likelihood of locating wide-ranging or rare species (Kimmel and Yahner 1991; Morrell et al. 1991; Yahner and Ross 1996; Kubel and Yahner 2007).

For example, we (Yahner and Ross 1995, 738) have noted that 46% of our contacts of wood thrush (*Hylocichla mustelina*) were unsolicited, but this song rate increased to 77% with the use of recordings. An increase in song rates or vocalizations would be advantageous to any effort when there are time or labor constraints.

Individuals know one another by use of duets. Duets are overlapping, coordinated bouts of sounds given by a mated pair or members of an extended family group; this phenomenon is known in 222 species of 44 bird families. In addition, individuals know one another by use of duets. Most duetting species are monogamous, tropical birds that jointly defend year-round territories; paired songs often only milliseconds apart.

Females of warbling antbirds (family Thamnophilidae), which are found in the tropics, sing more often and quicker when exposed to the playback of the song of another female song compared to playbacks of either a duet or solo male (Stutchbury and Morton 2008). This supports the idea that duets are intended to defend a mate rather than jointly defend a territory.

In other tropical birds (Stutchbury and Morton 2008), e.g., black-bellied wrens (*Thryothorus fasciatoventris*), they respond to a "duet code" that is reciprocal; for instance, if a female black-bellied wren sings song A, the male sings song W; but if song W is playbacked to the male, he will sing song A. This suggests that coupling of a duet code may serve to identify mates.

In some other species, like the marsh wrens (*Cistothorus palustris*) of temperate regions, vocal duels occur between rival males, and neighboring rivals try to match songs of each other (Gill 1990). Hence, duetting in marsh wrens may be a mechanism to show social dominance.

14.6 How Do Birds Identify Songs?

The question regarding how do bird identify their own song or those of others has been studied well (Gill 1990). As we might expect, cues used to identify a given song can vary with species, e.g., pure notes and pitch of the white-throated sparrow (*Zonotrichia albicollis*) are important to this species; if there are long intervals between notes or if notes waver in the song of a male white-throated sparrow, another male white-throated sparrow will not respond to this song of a conspecific. Other cues include syntax (this is the sequence of particular notes), e.g., brown thrasher distinguishes songs from the closely related gray catbird by two repetitions of each syllable. In addition, syntax is not important to indigo bunting; thus, the song of the indigo bunting is rhythmic timing, and there are frequent changes in individual notes. Thus, subtle changes in gestures of songs allow individuals to recognize others of the same species.

14.7 Song Repertoires in Birds

Songs in birds are considered extremely varied and vary like those of nonhuman primates (Gill 1990). Bird songs may range from a single song of white-throated sparrows to hundreds of songs found in some mockingbirds. Some birds, like winter wren (*Troglodytes troglodytes*), have a small repertoire, but their song lasts a long time (e.g., 8 s and consists of about 50 notes).

Some birds, such as the song sparrow, may sing a given pattern up to 12 times before switching to another (Gill 1990). What is the value of a large repertoire of songs in birds of a given species? Large repertoires may enhance (1) the attractiveness of a given male to a female, (2) permit a male to better compete with rivals, (3) discourage would-be territorial males (e.g., floaters), (4) simulate continued interest by listeners, or (5) indicate to females which males are older and more experienced. Taped song sequences >5 syllables played to common canaries (*Serinus canaria*) resulted in females building nests faster, laying the first egg sooner, and laying larger clutches compared to females exposed to song sequences with only 5 syllables. In great tits, males with large repertoires had young with a greater body size, which presumably enhanced reproductive success. Males with varied repertoires also may make it difficult for a neighboring male to assess who is the rival male and the distance away of the rival male.

14.8 Mimicry in Birds

Mockingbirds are known to imitate songs of dozens of sympatric birds; e.g., in Texas, imitate Bell's vireo (*Vireo bellii*) and great-tailed grackle (*Quiscalus mexicanus*). Migratory birds, such as the marsh warbler (*Aocephalus palustris*) of Europe, mimic birds encountered in migration or on the wintering grounds in Africa. This mimicry may tell a potential mate on the breeding grounds in Europe where a given territorial male may have overwintered in Africa. This may be important for a female seeking a mate that overwinters in the same area, thereby producing young with similar behavior.

Avian vocalizations, like a song, may be inherited, learned, or innovated (Gill 1990). Birds, such as chicken (*Gallus gallus*) and dove (family Columbidae), inherit calls even if raised in isolation; in some species, songs are inherited; but in the eastern meadowlark (*Sturnella magna*) calls are inherited, but songs are not. If any bird is raised in isolation, the song is less complex, but the song resembles that of the species.

There may be four stages to the development of a bird song (Gill 1990). In stage 1, also called the critical period, information about a song is stored for later use by a young bird; usually this stage is <1 year in duration. This stage gives a young bird the opportunity to hear the song of an older, experienced male on the breeding ground. The duration of stage 1 is only 60 days in the marsh wren, and 10–50 days in white-crowned sparrow (*Zonotrichia leucophrys*). Mockingbirds, however, add vocalizations throughout their lives. Stage 2 is termed the silent period, whereby syllables picked up in stage 1 are stored in the brain; stage 2 can last up 240 days in the swamp sparrow (*Melospiza georgiana*). Stage 3 is a practice period. It begins with a subsong in species that learn a song. A subsong is unstructured (it has been likened to an infant babbling). The fourth and last stage in the development of a bird song is the crystallization period. A bird needs to sing the song and get auditory feedback to perfect a song during this stage. In fact, for deafened birds, crystallization or perfection of the song does not occur.

Dialects are common in many species (Gill 1990). As mentioned earlier, sperm whales in the South Pacific have distinct dialects and swimming patterns (Fox 2003). We all have certain dialects; often we can get some clue as to the home of people by listening to their dialect. Dialects also are widespread in birds, and dialects likely involve learning. A question which arises is why do dialects occur in animals? Perhaps dialects parallel geographic speciation, in that isolated populations tend to develop a certain dialect. Perhaps an animal increases its reproductive success if it sounds like its neighbor. Else there may be an ecological hypothesis involved with the development of a dialect. Although doubtful, a dialect may mark the environment in which a young was successfully raised.

14.9 Sound-Producing Mechanisms

It is somewhat easier to determine the evolution of sound-producing structures than the evolution of displays because of the origin uses of these sound structures (Brown 1973). Mammals and birds typically use respiratory structures and the flow of air; in mammals, the larynx usually is the structure; air sacs, the syrinx, or chambers, often serve as resonating structure for sounds in frogs and toads (order Anura), howler monkeys (*Alouatta* sp.), swans (*Cygnus* sp.), etc. Other sound-producing substrates may involve inanimate objects such as a tree, the drumming by a woodpecker (family Picidae); or water, a beaver (*Castor canadensis*) slapping its tail on it. In some arthropods (phylum Arthropoda), an exoskeleton may be rubbed together, which is known as stridulation. Crickets (family Gryllidae), for instance, have auditory receptors that are more diverse than in vertebrates. Because the stridulation mechanism is adapted for very narrow limits, these receptors are tuned into the frequencies created by stridulation, making crickets tone deaf for other frequencies. Some moths also may be tone deaf, by hearing tones only between 16,000 and 100,000 Hz (frequencies per second), which is within the ultrasound range produced by their enemy, foraging bats. Moths have been known to change their flight course when they pick up high-frequency used by bats in echolocation, often 25,000–50,000 Hz. Often this flight change is a power dive to the ground.

Sound production in fishes tends to be of low frequency, and it seems to work best in the water when fish sounds are under 500 Hz (Brown 1973). Fish apparently use sound for various reasons: attraction of mates, defense against predators, threatening, or schooling. With underwater noise pollution on the rise, circumstantial evidence suggests that this may cause stress in freshwater fish and whales. For example, freshwater fish have doubled secretion of the stress hormone cortisol that is presumably caused by noises of high-speed boats; in marine waters ship noise or air-gun detonations has likely increased cortisol secretion in whales. Regardless of species, increased stress can be detrimental to growth, survival, and reproduction. In fact, noise in water is magnified five times that in the air.

Some recent studies have use underwater robots that hear calls of whales, including low-frequency of baleen whales (suborder Mysticeti), plus high-frequency calls of beaked whales (family Ziphiidae) (Anonymous 2007). This tracking of whale calls is important so that naval sonar operations on ships can avoid whales if whales are detected.

Off the coast of Australia, found that tourism vessels likely has increased; coinciding with tourism vessel increase has been a possibly 14% decline in bottlenose dolphin (*Tursiops* sp.) populations over the past 15 years. This calls into question the effects of low-level tourism vehicles versus use of research vehicles to monitor dolphin numbers. Underwater noises created by the engines of tourist vessels are louder than those of research vehicles. Although research vehicles also reduce populations, a decline in bottlenose dolphins has been nonsignificant, but gradual (9.5%) over 15 years. Thus, perhaps the disturbance caused by research vehicles is important and necessary because their use by researchers can detect future declines in populations of bottlenose dolphins and because research may enhance welfare of a population of whales via conservation of species.

14.10 Effects of Sound on Terrestrial Animals

The potential impact of noise on terrestrial animals is certainly of concern, e.g., there have been studies of the effect of artillery fire on red-cockaded woodpecker and the impact of noise created by logging trucks on northern goshawk (*Accipiter gentilis*) (Pater et al. 2009). Ears of frogs and toads can discriminate a wide range of frequencies, and chores by breeding frogs and toads in spring and summer, can record choruses by driving around at dusk and at night, are well-known and are an effective sampling technique (Hulse et al. 2001). However, vocalizations are known in only one salamander species, which is the Pacific giant salamander (*Dicamptodon ensatus*); this salamander gives faint squeaks to escape a predator. Sound production in reptiles is represented well by crocodiles (family Crocodylidae); these reptiles give roars and bellows during aggressive encounters. Rattlesnakes (subfamily Crotalinae) make alarm signals via rattles in tails; the eastern hognose snake (*Heterodon platirhinos*) makes a hissing sound that sounds like the rattle of a rattlesnake, which is an antipredator defense. Some geckos (family Gekkonidae) and some tortoises (family Testudinidae) are two additional groups of reptiles that produce sound. In the former group, the reason for sound production is unknown, but perhaps it has an antipredator function; in the later group, sounds are produced during courtship, which suggests a reproductive function.

Chapter 15
Ultrasounds and Other Types of Communication

15.1 Introduction

Humans potentially can hear between 20 and 20,000 Hz; with age, older human ears often lose part of this range (Yahner 2001). Sounds above 20,000 Hz typically are considered as ultrasounds, such as those of bats, which produce sounds between 25,000 and 140,000 Hz. The term echo refers to the fact that signals are sent out and they bounce back to the sender.

15.2 Echolocation in Bats and Birds

In 1793, bat biologists found that deafened bats were disoriented, but blind bats were not disoriented. Then in 1920, bat biologists believed that bats use echolocation at night to capture insects. But it took until 1930 when microphones were invented that evidence showed that bats use ultrasound to locate flying insects as prey. Today, the primary function of echolocation in bats is believed to be for food acquisition; a secondary function for bat echolocation is probably for orienting in darkness.

Why is echolocation in bats beyond the hearing range of humans (Yahner 2001)? Certainly, high-frequency use in bat echolocation is not to avoid detection by humans or other predators. Some bats are capable of flying about 30 km/h in relative darkness and capture an insect, which is the size of a dime and several meters away. These circumstances require a rapid series of signals about 30,000 Hz to match the size of a 11-mm insect; ideally, a foraging bat needs to emit a signal with a wavelength to match the size of the insect; by doing so, a bat probably gets a better echo back from prey for more precise information on location.

A bat, called the little brown myotis (*Myotis lucifugus*) (Fig. 15.1), uses many signals that enable it to detect subtle change in prey location. Their echolocation goes from about 25 pulses per second in normal hunting to about 200 pulses per second

Fig. 15.1 The little brown myotis, or little brown bat, is common in North America. As a winter strategy, this species migrates

(shorter duration) when a prey is detected and closing on to the prey. Hence, in the presence of prey, rate and duration of pulses change. How does a bat phase outgoing and incoming signals to detect an echo? Each species of bat has a fleshy projection in its ear, called the tragus, which acts as a reflecting surface for incoming signals and provides a second and slightly longer pathway for signals coming back to the bat. This second, phased signal enable bats to distinguish slight changes in the vertical positioning of objects in its path. Each bat species has a vocal signature in which it is <40,000 for the little brown myotis. Thus, bat detectors have been used to distinguish frequencies and of the features of bat calls to determine species present. But the use of bat detectors is not full-proof, with only about 70% accuracy.

Echolocation is not restricted to bats; it is found in two bird species, called the oilbird (*Steatornis caripensis*) (from South America) and the cave swiftlet (*Collocalia linchi*) (from Africa) (Gill 1990), use series of clicks in the 1,000–15,000 Hz range for discriminating objects in a dark cave. Because these clicks are within the hearing range of humans, they are not true ultrasounds and are only about 10% as effective in discriminating objects compared to discrimination in bats.

15.3 Echolocation in Shrews

Shrews (family Soricidae) are mouse-like small mammals that are also believed to echolocate. They are semifossorial, with poor eyesight. The northern short-tailed shrew (*Blarina brevicauda*) forages with its poor eyesight in leaf litter and moves quickly in dark, underground burrows and tunnels. This shrew produces sounds between 30,000 and 50,000 Hz; thus, echolocation in this species probably is effective only for 1 m or less; it uses echolocation to distinguish openings of 0.6 cm in diameter (e.g., burrow systems) and of objects (food versus nonfood).

15.4 Sounds in Pinnipeds, Toothed Whales, and African Elephants

Sound produced by pinnipeds (Seals, etc.) and toothed whales are typically within the range heard by humans (Yahner 2001). For example, the harbor seal (*Phoca vitulina*) produces two types of clicks around 7,500 Hz. In pinnipeds, functions of

15.4 Sounds in Pinnipeds, Toothed Whales, and African Elephants

Fig. 15.2 The African elephant is common to most people. Both male and female have tusks, whereas in the other well-known elephant, the Asiatic or Indian elephant, only males have tusks

underwater sounds seem to be more sophisticated than in bats or shrews. Underwater sounds by pinnipeds possibly are used to locate objects in water under poor light conditions, discriminate among objects (food versus nonfood), and communication. In toothed whales, sounds produced also are audible to humans, and the three functions found important to pinnipeds are also attributed to those of toothed whales, and a fourth function associated is to stun the prey. Sperm whales (*Physeter macrocephalus*), for instance, give clicks that can detect prey, like squid (order Teuthida), possibly up to distances of 750 m. Squid can swim up to 50 km/h, so to catch this prey, sperm whales stun them with high-intensity pulses to immobilize the prey until captured. This becomes effective in water because sound travels five times faster and creates up to 60 times the pressure than in air. Stranding by aquatic animals is caused likely by several factors, including those that may affect echolocation, e.g., slope of the shore, magnetism caused by iron in rocks along shoreline, effect of naval sonar on signals.

Low-frequency sounds, or infrasounds, are emitted by African elephants (*Loxodonta africana*) (Fig. 15.2). These animals produce sounds (14–35 Hz), which are audible to humans and sounds like distant thunder. Because these sounds are low in frequency, they are likely unimpeded by vegetation; also, these calls can be heard presumably for at least 4.0 km by elephants. Infrasounds in African elephants likely are used to attract mates and as dominance signals between rival males. The Elephant Listening Project, in the Central African Republic, perhaps has some practical implications if understood by humans. First, it may serve to estimate elephant densities. Second, it may be used to decipher when an elephant herd plans to raid a crop field. Third, this technique may be used in regional and locally to manage elephant populations. In Asia, the endangered Asian elephant (*Elephas maximus*) prefers nutritive crops over native vegetation; these elephants only raided crops if within the home range of elephants and raids were made only by larger males in groups of 1–3 (Williams et al. 2001). In Nairobi, the Save the Elephants program, found that playing loudspeakers with sounds of angry African bees (*Apis mellifera scutellata*) caused African elephants to flee, which enabled humans to control the elephants.

15.5 Introduction to Tactile and Electrical Communication

Many animals communicate by physically touching, which is obviously effective at short distances (Goodenough et al. 2001). Where an animal is touched may send very different signals; e.g., simple placement of the leg by a dominant dog on the back of a subordinate male sends a very specific signal of dominance. Honeybee scouts inform nest mates of the location of food by dancing; because the hive is dark, nest mates follow the movements of the dancer with their antennae.

15.6 Tactile Communication

Primates have taken tactile communication to a high level (Eibel-Eibesfeldt 1970). Primates in zoological gardens, e.g., chimpanzees (*Pan* spp.), need contact with others, including zookeepers; otherwise, they may deteriorate emotionally. Perhaps tactile communication in primates has roots with a younger animal needing to have contact with its mother. Also, chimps put their arms around each other when frightened; sometimes, a subordinate chimp may flee to be reassured and get a calming effect by the highest-ranking animal, as if to get protection by the dominant animal.

Tactile communication is certainly not confined to primates but is very important also in African elephants (Barnes 1984); in mother–young relationships, the mother continually touches and guides the young with her trunk; when two elephants meet, they often greet each other by touching the mouth of a conspecific with the tip of the trunk. Primates also take tactile communication to a higher level via grooming, which goes beyond simply removing parasites. In primates and birds (Gill 1990), self-grooming serves to care for the skin or feathers. But in social grooming by primates, very different things are occurring. Social grooming has two purposes (1) to establish and maintain pair bonds, and (2) to reduce tension. For instance, in chimpanzees, about 40% of the fights end in some form of reconciliation behavior, including outstretched arms, open hand gestures, eye contact, and kissing. In other animals, such as collared peccary (*Tayassu tajacu*) (Fig. 15.3) rub heads against the flanks and the rump of a conspecific, perhaps as a greeting ceremony. The female dik-dik (*Madoqua* spp.) of Africa rubs noses of her young, possibly as tactile reassurance.

Fig. 15.3 The collared peccary is pig-like, but it is not in the pig family. Peccaries are often called javelinas. Collared peccary represents a major prey of the large cat, the jaguar, where the two are sympatric (have overlapping distributions)

15.7 Play Behavior

Play is believed to be important in normal development of young in most mammals, some birds, and a few reptiles (Goodenough et al. 2001). Play typically involves tactile communication, anthropomorphically it is fun, but it may be difficult to define because it often involves bits and pieces of other behaviors, e.g., prey catching, mock fighting, and incomplete sequences, as with a young cat (*Felis catus*) pouncing on a leaf. In Canidae, play may include biting and shaking in young animals; domestic dogs will use play makers, such as a "bow," to indicate that biting and shaking would be viewed as part of play. There are at least three hypotheses given for play behavior (1) it gives an animal training for strength, endurance, muscular coordination for dominance hierarchies, defense of territories; (2) it allows animal to practice skills, e.g., grooming and sexual behavior; (3) it enables an animal to learn specific skills or improve perceptual abilities; e.g., it possibly has cognitive value.

15.8 Seismic Communication

Reptiles and amphibians are able to pick up vibrations from the ground, the water surface, or other substrate (Goodenough et al. 2001). Apparently, seismic signals travel farther than auditory signals. Did you ever feel the rumbling of the ground after an underground blast?

The female white-lipped frog (*Leptodactylus albilabris*) (Fig. 15.4) from Puerto Rican rainforest feels the thumps given by the male during courtship. Reptiles and amphibians likely use auditory communication to detect prey, but this function largely has not been undocumented scientifically (Goodenough et al. 1971). There is some evidence, however, that toads orient toward the calls of conspecifics, and geckos locate crickets by their sound. In kangaroo rats (family Heteromyidae), foot drumming, which presumably is seismically detected, enables these rats to defend a territory.

Fig. 15.4 The white-lipped tree frog, or giant tree frog, is native to Australia and is the largest known tree frog

Detection of seismic vibration is probably important to ground-dwelling snakes, salamanders (order Caudata), frogs, and caecilians (order Gymnophiona), which have no external ears. Some experiments suggest that detection of seismic vibrations in salamanders is twice as sensitive to that of frogs. Snakes probably send vibrations through the lower jaw via the quadrate-columella to the inner ear; also, although not as sensitive, snakes likely pick up seismic vibrations. In this group of animals, seismic activity is low frequency at 100–200 Hz, which pick up these frequencies through the skin; such frequencies may be produced by sounds of digging insects or approaching mammalian predators.

In water, tapping on the surface sends ripples across the water; female water striders (*Rhagadotarus* spp.) receives "vibes" created by a patterned series of ripples before mating; male also sends out other patterns that signal territorial behavior. Thus, this complex means of communication, sometimes called ripple communication, perhaps has some origin in males using this motion to aerate eggs.

Lateral lines also are found in aquatic vertebrates; water currents can be detected, although there is some controversy about whether these animals respond to sound as a natural stimulus itself or to the pressure created by the sound (Goodenough et al. 2001). Lateral lines may be used by some fishes and amphibians to detect surface waves created by movements of prey, such as aquatic insects, on the surface of the water. Perhaps lateral lines facilitate schooling, e.g., herrings (*Clupea* spp.) have lateral lines only on the head and not on flanks, which may be sensitive to turbulence in water in head created by conspecifics.

15.9 Electrical Communication

Air does not conduct electricity, yet water does; moreover, freshwater does not conduct electricity as well as saltwater (Goodenough et al. 2001). Thus, many fishes, e.g., sharks (superorder Selachimorpha), have structures in heads and rays, termed ampullae of Lorenzini (Pough et al. 2002). In sharks, this electric sensitivity is used to detect prey, with muscles of all prey producing slight electrical discharges (Goodenough et al. 2001). Therefore, sharks even can detect hidden prey. Some sharks, such as the hammerhead sharks (*Sphyrna* spp.), also use the electromagnetic field of Earth to navigate and migrate. Electricity can also be used as a weapon by some organisms, e.g., torpedo ray (*Torpedo* spp.) of the Mediterranean or electric eel (*Electrophorus electricus*) of South America can produce enough voltage for hunting and as defense. Some fishes produce very small voltages in brief pulses (Pough et al. 2002), which can be used to distinguish individuals, ripe females, or sexually active males in certain species found in visually impaired (e.g., murky) waters; hence, electrical communication may be used for communication.

Mammals termed monotremes (only three extant species; order Monotremata), include the duck-billed platypus and echidnas (spiny anteaters); these mammals use electroreception to detect prey (Vaughan et al. 2000; Goodenough et al. 2001). In the platypus, the bill is packed with dense electroreceptor cells that can pick up weak

15.9 Electrical Communication

electrical currents given off by prey while this animal feeds at night in the bottom of a murky stream. Of the two species of echidnas, the Australian (not the New Guinea) has a slender, beak-like rostrum with electroreceptors.

These are not the only mammals with electoreceptors (Vaughan et al. 2000). The star-nosed mole (*Condylura cristata*) of North America, for instance, is an aquatic mole of our region with 22 fleshy appendages off its snout; each appendage is covered with touch receptors, called Eimer's organs, that communicate with the brain if prey is touched in stream bottoms, begin this mammal about 25,000 of these receptors, which is about five times that in the human hand. Thus, the nose of this animal is primarily for tactile purposes and not for olfaction, but there is some evidence that the 22 appendages also are used as electroreceptors.

Chapter 16
Winter Strategies

16.1 Introduction to Winter Strategies

Animals have three options to survive the harshness of winter: stay active, hibernate or undergo torpor, or migrate. Dormancy has occasionally been used to include both hibernation and torpor (Morse 1980). Both ectotherms and endotherms become dormant in response to changes in some optimal environmental condition, which presumably results in substantial energy savings. Some species, such as some bats, e.g., red bat (*Lasiurus borealis*), may use more than one strategy (dormancy and migration) to deal with environmental harshness. By adopting one or more strategies, a species can survive environmental harshness without evolving a flexible foraging repertoire.

Regardless of developing a dormancy strategy, three problems arise, including a need to obtain adequate food for survival with decreased metabolic rates, to find an appropriate site for dormancy, and to obtain adequate food once dormancy is over. But first, we need some definitions. Hibernation, as in the woodchuck (*Marmota monax*), is an extended state of dormancy, whereby body temperatures and other physiological functions decline dramatically for several weeks in mammals, is restricted to animals about woodchuck size or smaller. This is likely because a larger animal would have to store too much body fat to survive this time period. As we will see, food in a hibernating animal is stored as appreciable body fat. Also, in large animal it would require considerable energy to warm up once aroused from hibernation. Winter lethargy, as in the black bear, has an extended semistate of inactivity, whereby body temperatures and other physiological functions decline somewhat for several weeks. In the black bear, food resource is stored as body fat. Torpor, as exemplified by the eastern chipmunk, is a short-term, shallow state of inactivity.

16.2 Hibernation

The desert tortoise (*Gopherus agassizii*) (Fig. 16.1) is the largest ectotherm in the North American deserts, with some having a shell length of 1 m. When environmental temperatures lower around November each year and plant food becomes scarce, the desert tortoise goes into hibernation (Pough et al. 2002). Some ground squirrels (*Spermophilus* spp.), e.g., Richardson's ground squirrel (*Spermophilus richardsonii*) may go into hibernation but arouse periodically. The Richardson's ground squirrel has a very short activity season, with adults going into hibernation in mid-July and emerging in mid-March. This ground squirrel slowly enters hibernation and has periods of torpor alternating with periods of arousal. Temperatures become around 0°C by late December in the burrow system and body temperatures go from 37–38 to 3–4°C. To survive winter, this ground squirrel lives on stored body fat. Not all animals in the same environment adopt the same strategy to survive winter. For instance, the pika (*Ochotona princeps*) is a diurnal species may be sympatric with ground squirrels, being found in the mountainous grasslands of western North America. The pika does not hibernate with the onset of winter, but rather stays active and feeds on stored grass, which it caches in its burrow system among rocks.

16.3 Hibernation in the Woodchuck

Probably the most famous hibernating animal is Punxsutawney Phil, which is a woodchuck that is awakened from hibernation on February 2 to give an annual weather prediction (Yahner 2001). The length of hibernation varies with latitude; for example, the duration of hibernation may extend from early November to early February or March in Pennsylvania but from mid-September to late March or early April in southern Canada. When woodchucks emerge from burrow systems in late winter or early spring, they do so infrequently and limit aboveground activity limited to 1 h or less. Mating occurs soon after emergence, which is probably the origin of Groundhog Day.

Hibernation in woodchucks is very energetically efficient, saving about seven times the amount of energy that is required to stay active (Yahner 2001). In autumn, a woodchuck gains about 30–40% more body weight as body fat, then loses this

Fig. 16.1 The desert tortoise is an ectotherm native to deserts of southwestern USA and northern Mexico (Mojave and Sonoran)

weight. Imagine weighing 150 pounds as a human in summer and gaining 50–60 pounds in autumn every year. The heartbeat of a hibernating woodchuck goes from 100 beats per minute to 15 per minute and body temperature goes from 38 to 8°C.

16.4 Hibernation in Bats

Some species of bats undergo hibernation in winter [e.g., little brown myotis (*Myotis lucifugus*)], whereas the sympatric red bat is migratory. The length of hibernation varies with species, being 6–8 months in little brown myotis. This bat lowers its body temperature to about 1°C above ambient temperatures, where is often hibernates deep in caves in large clusters, which sometimes may number 1,000–6,000 individuals per cave. Hibernating bats in caves need protection; if handled or aroused by a disturbance, individuals can have tremendous energy loss.

16.5 Winter Lethargy in Black Bear

Winter lethargy in bears is a metabolic and an ecological wonder (Yahner 2001). Black bears begin entering dens to undergo winter lethargy when body fat reserves are accumulated and food becomes scarce in fall. But the timing varies with latitude, varying from October to early January. Adult females tend to enter lethargy earlier than younger animals or adult males.

Entrance into winter lethargy by black bears is not all-or-none; instead, bears gradually adapt behaviorally and physiologically prior to entrance over about a 1-month period (Yahner 2001). Emergence from winter lethargy seems to be triggered by spring warming in mid-March to early May. Like woodchucks, bears rely on fat reserves as an energy source over winter; however, needs for winter energy are lower than those needed by woodchucks. In fact, black bear gain only 20–27% of their body weight as fat. Metabolic changes also are not as dramatic; for instance change in body temperature only goes from 40 to 30°C. A heart rate of 30–40 beats/min declines to 8–10 beats/min. Oxygen consumption in bears is about 50–60% of normal consumption. Where black bears become a metabolic wonder is when these lowered metabolic processes occur while a lethargic black bear does not eats, drinks, defecates, or urinates. In addition, an adult female may give birth and lactate at this time.

Two reasons have been given to explain the evolution of winter lethargy in bears, which include food scarcity and the need to care for young (Yahner 2001). The second factor seems more plausible, because in southern latitudes, even with food available, females go into winter lethargy, but males do not. Young bears (cubs) are born very small, weighing less than 1% of the weight of the mother. Thus, very small young in bear have tremendous heat loss because of a high surface area to volume ratio, so heat from mother is vital for survival. A female bear is physically capable of being aroused to defend her young, such that disturbing female in winter lethargy to defend her young can cause a 50% energy drain, thereby affect young survival with reduced lactation as a result.

16.6 Torpor in Eastern Chipmunks

Eastern chipmunks undergo torpor (Yahner 2001). When chipmunks go into torpor, they have very little weight gain; instead, food for winter use is larder hoarded by chipmunks in an underground burrow system. In some chipmunks, torpor may last one day at a time or never occur, whereas torpor may extend several days, depending on the individual. When a chipmunk awakens from torpor, they feed on food in the larder. It is probably adaptive for chipmunks to undergo torpor, thereby lowering body temperatures, heart rates, breathing rates, and, hence, the need for a chipmunk to store as much food in the cache.

16.7 Daily Torpor in Aerial Animals

Perhaps daily torpor is more common in aerial animals compared to terrestrial animals (Yahner 2001). Thus, daily torpor is perhaps best related to the demands of both endothermy and flight. For example, in winter, black-capped chickadees can lower their body temperatures overnight from 40–42 to 29–30°C, which results in a 30% reduction in energy use (Pough et al. 2002). Hence, black-capped chickadees rely on fat reserves stored during the day. Daily torpor is necessary in these types of animals because by not lowering their body temperatures, a black-capped chickadee would have to add 0.92 g of fat, but only store 0.80 g of fat; thus, a black-capped chickadee would otherwise starve overnight without daily torpor. These birds forage immediately at sunrise no matter how bad the weather is to replenish energy stores. Ruby-throated hummingbirds (Fig. 16.2), when in daily torpor overnight, are unresponsive to lost stimuli and incapable of normal activity (Gill 1990). In this bird,

Fig. 16.2 The ruby-throated hummingbird is solitary and migratory, and it has a breeding range throughout most of eastern North America

16.7 Daily Torpor in Aerial Animals

oxygen consumption drops by 75%, and body temperature declines from 32 to 20°C, resulting in a 27% energy savings.

Vespertilionid bats (family Vesperitionidad) undergo daily torpor overnight in summer when ambient temperatures drop (Vaughan et al. 2000). At 35°C compared to 5°C, energy savings was 33% because of torpor; thus, in interest of saving energy, bats reduce flight time as much as possible. Some species might conserve energy by huddling together in dens (Vaughan et al. 2000) or orienting body or body parts (e.g., wings) toward solar radiation (Gill 1990).

The length of torpor overnight may be in tune with the amount of energy stores (Pough et al. 2002). For instance, ruby-throated hummingbirds that had a 12% reduction in foraging time because of bad weather and went into torpor overnight for 2 h; in contrast, birds losing 21% foraging time, went into torpor for 3.5 h the following night.

Chapter 17
Migration, Orientation, and Navigation

17.1 Introduction

Storing food, becoming dormant, or migrating is one of three strategies that an animal may adopt to survive environmental harshness, e.g., winter cold and snow. The annual appearances and disappearances of birds were long a mystery (Gill 1990). For instance, Aristotle knew that cranes migrated from Asia to the Nile, but he believed that smaller birds hibernated. We now know that over 180 bird species migrate from Europe and Asia to Africa, and that over 170 species migrate from North America to the tropics each year.

17.2 Advantage of Migration Over Dormancy

When an animal migrates, it may exploit different feeding opportunities in a favorable climate, which also may include less pressure for food shortages and predation risks (Goodenough et al. 1701). However, migration may be risky, with only about 50% or less of migrating songbirds and waterfowl that migrate return the following year. Three general benefits may be increased reproduction, reduced competition, and reduced predation.

A reproductive benefit may be accrued by breeding in an area with longer days and high food abundance, e.g., in northern latitudes for songbirds or in warmer coastal bays where waters may be warmer for calves of gray whales (Goodenough et al. 2001). In the tropics, with a greater number of species, reduction in competition conceivably is a benefit to a species migrating northward in spring. A reduction in predation is a benefit given often for there being more migratory ungulate species than nonmigratory ungulate species; also, when birds restrict breeding to short time period in northern areas (far north), less time is available to predators to exploit this prey.

Disadvantages of migration may include it being costly in terms of energetics or risk (Gill 1990). For instance, migration in warblers (family Parulidae) to

Central America is like a human running a 4-min mile for 80 straight hours. It takes 6–8 times more energy to migrate than to rest in birds. The monarch butterfly is very sensitive to freezing temperatures; one cold night, over two million on the Mexican wintering grounds were killed (Goodenough et al. 2001).

17.3 Distances of Migration

Why do migrations distances vary among species? For instance, gray whales (*Eschrichtius robustus*) make annual migrations of about 10,000 km from a northern Pacific feeding area to breeding grounds in Mexico (Goodenough et al. 2001). Hence, these long-distant migrations makes the scale difficult to protect this species (e.g., to preserve wilderness preservation in oceans; Sloan 2002). Migratory distances in terrestrial mammals typically are shorter than for aerial or aquatic mammals (Goodenough et al. 2001). Serengeti wildebeest (*Connochaetes taurinus*) (Fig. 17.1) travel about 1,700 km in migration; bats may travel from several 100 km to over 1,500 km, depending on the species. The sanderlings (*Crocethia alba*), which is a sandpiper, migrates from Chile in winter to breeding grounds in Arctic for a distance of 7,500 km, which spans 230 h. The northern elephant seal travels about 21,000 km from foraging to breeding areas in the northern Pacific to the Channel Islands of California.

Migration is different from nomadic wanderings, which are common when food is unpredictably sporadic and irruptive [e.g., pine seeds exploited by red crossbills (*Loxia curvirostra*)] (Goodenough et al. 2001). In contrast, migration is in response to predictable increases in food, e.g., seasonal abundance of forest insects in eastern forests.

Besides latitudinal migrations, some species migrate altitudinally, e.g., mountain sheep (*Ovis* spp.) (Fig. 17.2) in the West (Morse 1980). In western Wyoming, mule deer and pronghorn migrate 20–158 km (12–100 miles) and 116–258 km (72–160 miles), respectively; because of this migration, these corridors need protection from humans

Fig. 17.1 The wildebeest, or gnu, is a native to Africa. They are grassland ungulates (animals with hooves); they are well-known for their annual migrations to fresh grass, especially across the Serengeti plains of the national park

17.5 Timing of Migration

Fig. 17.2 Mountain sheep, or bighorn sheep, have males with big curving horns. Bighorn sheep are native to North America, having crossed into this continent via the Bering Land Bridge from Siberia

(e.g., housing developments). In white-tailed deer of northern latitudes, deer may yard or migrate to an area that provide shelter during winter from cold and wind (deer book 135).

17.4 Energy Stores for Migration

Migratory animals typically use body fat, which produces twice the energy and water than carbohydrates or protein per unit amount (Morse 1980). The migratory blackpoll warbler, for example, consumes fat by eating energy-rich food just before migration. A blackpool warbler may go from a body weight of 11 g to about 21 g. By comparison, normal body fat weight gain may only be 3–5% in nonmigrating species. On average, a migratory bird loses 0.9% of its body weight per hour of flight. Thus, the amount of body fat determines flight ranges; small birds with fat reserves of 40% of their body weight can fly about 100 h and go 2,500 km.

17.5 Timing of Migration

Compared to our airlines, arrival and departure in birds are very precise. For instance, cliff swallows (*Petrochelidon pyrrhonota*) of *the* San Juan Capistrano mission in California arrive precisely each year on 19 March. Migratory birds exhibit restlessness or Zugunruhe, which presumably is linked to hormones near the migratory seasons; restlessness is not seen in nonmigratory species. In winter prior to spring migration, restlessness may be linked to increased length of light during a day, excess feeding, fat deposition, and weight gain; in fall before fall migration, restlessness is linked to a decrease in the amount of light per day. Geese and American robins move north with the spring thaw, along regions with a mean ambient temperature of 2°C (2°C isotherm).

A study conducted in Maine showed that most birds (60/105 or over 50%) showed a weak but a gradual change to arrival dates in spring. Arrival dates were based on 12 years of data beginning in 1994. Five species, in particular, e.g., warbling vireos (*Vireo gilvus*) had pronounced changes in arrival times. The study concluded that global warming was a possible factor in shifts in arrival dates, but that dates may be driven by photoperiod or some other environmental change. Favorable weather conditions also stimulate departure. For example, strong northwest winds create a barometric depression on the east coast of the USA, which is favorable to raptors migrating along Hawk Mountain, which is near Kempton, Pennsylvania.

In some species, males migrate north before females, as with male red-winged blackbirds that arrive early to establish territories; in some other species, young migrate south before adults; e.g., least flycatcher (*Empidonax minimus*). It has been shown that males migrate north in spring about 5–8 days before females in Wilson's warbler (Benson et al. 2006). In this warbler, immatures migrate earlier sooner than adults (about 13 days earlier) in autumn to wintering areas in Central and South America. Perhaps a delay in adult migration is because adult must undergo prebasic molt (feather replacement) before migration.

In black brant (*Branta bernicla*), older birds arrive in their breeding grounds (Alaska from Baja California) and stop over for shorter time period when in route than younger birds when in route. Reasons for time to get to the breeding ground may include that fitness in adults is maximized if older because younger birds conserve energy by taking longer to migrate because younger birds are unlikely to mate and adults may be better foragers at stopover points than younger birds.

A unique case among migratory songbirds may occur in white-throated sparrows. In this species, there are about equal numbers of two morphs: white-striped and tan-striped in both sexes; 95% of pairs have on bird of each morph. White-striped males are more aggressive, have higher rates of attempted polygyny, higher rates of intrusion into neighboring territories, but lower parental care and mate guarding. In spring, males arrive on territories before females, but white-striped females arrive 1.3 days earlier than tan-striped females, which may afford advantages to white-striped females.

About 46% of birds return to area used for breeding in previous year, e.g., Swainson's or olive-backed thrush (*Catharus ustulatus*). Perhaps lower return rates occur in species lacking geographic differences in vocalizations (dialects); for example, 5% in Baird's sparrow (*Ammodramus bairdii*).

Some species migrate during the day, e.g., hawks, taking advantage of warm rising air currents; most small landbirds, e.g., warblers, migrate during the night, which would reduce predation, plus migrating at night gives landbirds about 12 h of daylight to feed. Shorebirds migrate at several kilometers, but ducks and landbirds typically under 5,000 km. Why bird migrate at various heights in unknown.

17.6 Navigational Routes

In North America, bird migration mainly oriented north–south, following coasts and mountain ranges, perhaps because mountain ranges are oriented north–south (Goodenough et al. 2001). In the Old World, mountains and other features are east–west, which corresponds to the direction of migration.

Some animals may get around in a random fashion; desert ants (*Cataglyphis bicolor*) wander in any direction when seeking food (up to 100 m from its nest); apparently, these ants know how many turns and steps it takes, seemingly using a memory snapshot. If a researcher places an ant at a distant site, it does not know how to get back to the nest.

17.7 Cues Used During Navigation

Besides the possible use of memory snapshots in desert ants, other species probably use a variety of cues to navigate (e.g., home or migrate) (Goodenough et al. 2001). Some birds return to the exact place if moved from the area; For instance, homing pigeons can return from distances of 800 km. These birds use the sun compass and perhaps the Earth's magnetic field. If the homing pigeon and other bird species, like the European starling, rely on the sun as a compass, an unanswered question is how do these species compensate for the 15° in the position of the sun relative to the Earth per hour or how do these species know that the sun rises in the East.

Visual cues, such as mountains and shorelines are important to diurnal migrants and short-distant movements (Goodenough et al. 2001). Digger wasps (*Sphex* spp.) use landmarks to locate their nests. However, animals that navigate or migrate at night may use the stars. Migrating bird do not use the North Star, in part, possibly because this star is stationary in the night sky, but rather they may use constellations with 35° of the North Star. But how do birds navigate at night when the sky is overcast? This is where magnetism from the Earth becomes important.

There probably are three aspects to magnetism produced by the Earth: polarity, lines of force, and intensity (Goodenough et al. 2001). The aspect important to the navigation of a species seems to vary. For instance, polarity, i.e., positive at the North pole but negative at the South pole, seems to be used in navigation by lobsters (family Nephropidae) and bobolink (*Dolichonyx oryzivorus*). Lines of force, i.e., lines parallel to the equator but perpendicular at the poles, apparently are important to green turtles and other sea turtles (superfamily Chelonioidea). In contrast, small differences in the intensity of the field may be used by homing pigeons and American alligators.

Olfaction cannot be discounted for orientation (Goodenough et al. 2001). Salmon, for instance hatch in cold, freshwater of rivers or lakes and then swim via streams to the sea. Depending on the species, salmon may spend 1–5 years in the sea until reaching breeding condition. Perhaps salmon use an olfaction (chemical trail) to a stream of birth; salmon may learn this scent soon after birth. There are two theories for the source of the steam odor. First, it may be a combination of the scent given by rocks, soil, and plants in the stream. Second, young salmon may learn the smell of pheromones given off by conspecifics.

17.8 Learning to Navigate

Navigational abilities in birds appear to be partly innate and partly based on experience (Goodenough et al. 2001). Evidence for this comes from the fact that young migrants typically are lost more often than older, experienced birds. In many cases, a young bird often loses the migratory route. In short, a young bird is more likely to become lost than an older bird during migration and become a rare visitor to an area.

Chapter 18
Competition

18.1 Introduction

For all or most in the wildlife, resources, e.g., food, space, or mates, are in short supply sometime during the year. A shortage of a resource can lead to interspecific or intraspecific competition (Morse 1970). We might place competition into either of two types: interference and exploitative. Interference competition typically involves activities that directly or indirectly limits access to a resource by a competitor, usually via aggression, e.g., actual fights. Exploitative competition, on the other hand, may occur when use of a resource is denied because another organism use the resource first, e.g., use of a den site. Why should a species compete for a resource? It may risk injury. Furthermore, competition may divert time and energy away from victor, it perhaps by exhaustion may make a victor more vulnerable to predation, or, in the case of male–male combat, combat itself may attract predators.

Interference competition may exist in the wild, but it may be difficult to observe for two reasons (Yahner 2001). Once a winner occurs in the wild, it may be unlikely to occur again. Also, interference competition may have been prominent evolutionarily; hence, it may have been "weeded" out much earlier by evolution. For one or both reasons, we may seldom see it occurring in the wild.

Mates often are a major resource for which there is competition if for mates (Goodenough et al. 2001). As a consequence, males often have evolved elaborate coloration in diurnal species, displays, or structure that may aid a male in male–male combat. Direct contact, for example, as seen in macaques (*Macaca* spp.) (Fig. 18.1), occurs when subadult males confront adult males in the presence of receptive females. As a consequence of in this scenario, subadult males often are injured severely by adult males.

Fig. 18.1 Besides humans, the macaques are very widespread, ranging from Asia to Africa, with as many as 22 species. The best known macaque is the Rhesus macaque or monkey

18.2 Interspecific Competition

Vandalism is documented known in house wrens (*Troglodytes aedon*), whereby house wren removes eggs and destroys nests of many bird species; this wren is believed to be responsible for declines in the Appalachian Bewick's wren (*Thryomanes bewickii*) until the 1930s, which was once widespread in the eastern USA (Yahner 2000). Interspecific competition has been suspected in two weasel (family Mustelidae) species: northern river otters (*Lontra canadensis*) and mink (*Mustela vison*) (Yahner 2001). Both weasels are adapted to aquatic environments, with the otter having a cylindrical body, a dorso-ventrally flattened tail for swimming, waterproof fur, and webbed feet. In contrast, the mink has an elongated body, semiwebbed feet, and waterproof fur. But there is temporal and spatial segregation of activity in these two weasel species, with northern river otters foraging in solely in water and at any time of the day, whereas mink forage along the shoreline and at night. Moreover, interspecific competition is minimized in otters and mink because of food selection. Otters are bigger (5–13 kg) than mink (1–2 kg), so otters select larger prey. At least 93–100% of the diet of otter is mainly slow-moving and bottom-dwelling species, but the smaller mink feed on only 7–59%; many of the fish are very small, and the diet of mink includes other things found along the edge of the water, e.g., eggs of birds. In short, interspecific competition is negligible between these two aquatic weasels.

Black vultures and turkey vultures are scavengers and use communal wintering roosts together; as mentioned earlier, turkey vulture has an excellent sense of olfaction; when the turkey vulture locates food, and the more aggressive black vulture takes over by capitalizing on the food (Wright et al. 1986). We might view this as interspecific and interference competition.

Is there always competition for food between sympatric species of similar size (Marti and Kochert 1995)? As an example, both red-tailed hawks (*Buteo jamaicensis*)

Fig. 18.2 The Sika deer, the Japanese spotted deer, is native to Asia and has been introduced into the USA and other countries. The Sika deer retains its spots throughout its lifetime

and great horned owls are generalist feeders, and there is about a 50% overlap in food resources used. Perhaps diet is separated because the red-tailed hawks prefer reptiles as food, whereas the great horned owls feed heavily on invertebrates.

Interspecific competition potentially may occur between white-tailed deer and snowshoe hares (a species of special concern in Pennsylvania) (Scott and Yahner 1989). Both species fed on the same trees, but deer browsed heavily on red (*Acer rubrum*) and sugar maple (*Acer saccharum*), whereas hares browsed heavily on striped maple (*Acer pensylvanicum*), blackberry (*Rubus alleghenensis*), and yellow birch (*Betula lenta*).

Exotic species have caused interspecific competition. For instance, Sika deer (*Cervus nippon*) (Fig. 18.2) are native to Japan, endangered in Japan, and are the national emblem of that country. Sika deer are about two-third the body size of white-tailed deer (Sika deer are not the Key deer; the latter are only about one-half body size of Sika deer). From a behavioral perspective, Sika deer have broader and more diverse diets than other deer, in part, perhaps because of their smaller body size. Sika deer are known to displace red deer (elk) in New Zealand; Sika deer also are known to outcompete white-tailed deer in enclosures in Texas. Moreover, in Maryland, Sika deer may be displacing white-tailed deer. Cane toads (*Bufo marinus*), which were introduced in 1935 into Australia, had no natural predators. Cane toads are native to Central and South America; because of their population explosion and lack of predator, they outcompeted native species for food in the recent past. Exotic ants (yellow crazy ants, *Anoplolepis gracilipes*) affected rainforest birds on Christmas Island, where it was introduced. This ant presumably has altered the foraging behavior of island thrushes (*Turdus poliocephalus*); perhaps more importantly, these thrushes showed lower nesting success and juvenile counts lower in ant-invaded forests (Davis et al. 2008); counts of emerald dove (*Chalcophaps indica*) also exhibited 9–14 less population abundance in ant-invaded forests. As a last example, exotic pigs (*Sus scrofa*) is one of eight species of pigs worldwide

(Yahner 2001). Feral pig populations are actually the Eurasian wild boar gone wild. The Eurasian wild boar was introduced by hunters around the turn of the twentieth century, or it escaped from farms. In the USA, pigs were left to free roam in forests around farms. There are four major loci of feral pigs today: in southeastern USA, in California, on eight islands in Hawaii, and in Puerto Rico/Virgin Islands. Currently, feral pigs exist in at least 23 states. Because of the foraging behavior of feral pigs, whereby the forest floor is uprooted, and their diet (anything from bird eggs, salamanders, to acorns), feral pigs likely have a major interspecific competitive impact on other acorn-dependent species in the eastern forest and elsewhere. Feral pigs also potentially transmit diseases, e.g., trichinosis, to domestic pigs and humans. Eradication of an exotic species can be costly. For example: to capture 200 pigs takes about 3,489 h (or 68 h per pig) and costs can be USD 623,601 (USD 3,118 per pig in 2008). Trapping seems to be most successful (>70% of individuals), although hunting and use of Judas pigs also have some value. Judas pigs are radio-collared pigs that are used to locate groups of feral pigs, which are difficult to find by other methods. Judas pigs are released and later they join social units of other pigs, which makes these units easier to find and to eradicate (McCann and Garcelon 2008).

Interspecific competition often may not involve an exotic species. In the past, there was concern that interspecific competition may exist between native elk and white-tailed deer. Competition may exist to some extent in winter for browse; but in growing season, grasses and forbs are more important to elk (Collins et al. 1978). Elk feed on browse, 1–35%; grass, 45–75%; forbs 10–50%; whereas white-tailed deer rely on browse, 55–75%; grass 15–25%; forbs, 5–30% on a year-round basis.

Black bears in southern latitudes tend to den (used as hibernacula) in tree cavities (Johnson and Pelton 1981; Alt 1984). In the Great Smoky Mountains National Park dens of black bears averaged 11 m above ground, which protect black bears from predators (humans and dogs) and inclement weather (floods). If a bear in the south gets wet from winter floods, the bear can lose 5–20% more body heat than a dry bear; denning in a tree also gives about 15% more insulation than no tree den at all. If this is the case, timber management practices in the South need to leave large trees with cavities for black bears. In contrast, in more northerly regions, bears den at ground level or about 1 m below level (under uprooted tree). In the North, dens of black bears covered with snow (or bears covered with snow) save have a 27% savings of energy. Only about 5% of bear dens are lost annually to flooding. But only 5–9% of northern dens of black bears are reused each year, perhaps because predators may learn the location of ground dens or disease transmission is reduced.

As another example, consider the gray and fox squirrel. A potential exists for exploitative competition between sympatric gray and fox squirrels for home sites (Edwards and Guynn 1995). However, little overlap occurs between these two species in the use of tree cavities or leaf nests. Gray squirrels more readily use tree cavities compared to fox squirrels, and leaf nests of gray squirrels are constructed closer to ground level and in smaller trees than fox squirrels.

Competition of any sort, need not be limited to animals, but may occur also in plants. For example, the exotic amur honeysuckle (*Lonicera maackii*) is presumably outcompeting native trees in Ohio for light or perhaps nutrients.

References

Allee WC (1938) The social life of animals. Beacon, Boston
Alt GK (1984) Black bear cub mortality due to flooding of natal dens. J Wildl Manage 48:1432–1434
Armitage KB (1998) Reproductive strategies of yellow-bellied marmots: energy conservation and differences between the sexes. J Mammal 79:385–393
Baker K (1990) Toad aggregations under street lamps. Br Herpetol Soc Bull 31:26–27
Balcom BJ, Yahner RH (1996) Microhabitat and landscape characteristics associated with the threatened Allegheny woodrat (*Neotoma magister*). Conserv Biol 10:515–525
Barash DP (1977) Sociobiology and behavior. Elsevier, Amsterdam, 378 pp
Barrette C (1977) Fighting behavior in muntjac and the evolution of antlers. Evolution 31:166–179
Barry RE Jr, Francq EN (1982) Illumination preference and visual orientation or wild-reared mice, *Peromyscus leucopus*. Anim Behav 30:339–344
Bauer GB (2005) Research training for releasable animals. Conserv Biol 19:1779–1789
Baxter RJ, Flinders JT, Mitchell DL (2008) Survival, movements, and reproduction of translocated greater sage-grouse in Strawberry Valley, Utah. J Wildl Manage 72:179–186
Beier P (1991) Cougar attacks on humans in the United States and Canada. Wildl Soc Bull 19:403–412
Beier P (1995) Dispersal of juvenile cougars in fragmented habitat. J Wildl Manage 59:228–237
Beier P (2007) Learning like a mountain: lessons on conserving habitat corridors. Wildl Prof 1(4):26–29
Berger J (2004) The last mile: how to sustain long-distance migration in mammals. Conserv Biol 18:320–331
Berger J (2007) Fear, human shields and the redistribution of prey and predators in protected areas. Biol Lett 3(6):620–623
Bertram CRB (1978) Living in groups. In: Krebs JR, Davies NB (eds) Behavioural ecology: an evolutionary approach. Sinauer Associates, Inc, Sunderland, MA, pp 64–96
Bertram CRB (1984) Blood relatives. In: MacDonald D (ed) The encyclopedia of mammals. Facts on File Publications, New York, pp 34–35
Blumstein DT, Fernández-Juricic E (2004) The emergence of conservation behavior. Conserv Biol 5:1175–1177
Bolen WL, Robinson EG (1995) Wildlife ecology and management, 3rd edn. Prentice Hall, Inc, Englewood Cliffs, NJ
Bollinger EK, Linder ET (1994) Reproductive success of Neotropical migrants in a fragmented Illinois forest. Wilson Bull 106:46–54
Bowne DR, Bowers MA, Hines JE (2006) Connectivity in an agricultural landscape as reflected by interpond movement of a freshwater turtle. Conserv Biol 20:780–791

Bradbury J (1977) Lek mating behavior in the hammer-headed bat. Zeit Fur Tierpsychol 45: 225–255
Bramble WC, Yahner RH, Byrnes WR (1992) Breeding-bird populations following right-of-way maintenance treatment. J Arboric 18:23–32
Bratton SP (1980) Impacts of white-tailed deer on the vegetation of Cades Cove, Great Smoky Mountains National Park. Proc Ann Conf Southeast Assoc Fish Wildl Agencies 33:305–312
Brotherton PNM, Rhodes A (1996) Monogamy without biparental care in a dwarf antelope. Proc Royal Soc Lond B 263:23–29
Brown JL (1973) The evolution of behavior. W.W. Norton & Company, Inc, New York, 76 pp
Campbell JM (1993) Effects of grazing by white-tailed deer on a population of Lithospermum caroliniense at Presque Isle. J PA Acad Sci 67:103–108
Caro T (2005) Antipredator defenses in birds and mammals. The University of Chicago Press, Chicago, IL
Chapman RC (1978) Decimation of a wolf pack in arctic Alaska. Science 201:365–367
Cheng R-C, Tso I-M (2007) Signaling by decorating webs: luring prey or deterring predators? Behav Ecol 18:1085–1091
Clutton-Brock TH, O'Rianin MJ, Brotherton PNM, Gaynor D, Kansky R, Griffin AS, Manser M (1999) Selfish sentinels in cooperative mammals. Science 284:1640–1644
Compton JA (2007) Ecology of common raccoons (*Procyon lotor*) in western Pennsylvania as related to an oral rabies vaccination program. Dissertation, The Pennsylvania State University, University Park, PA
Conover MR (1997) Monetary and intangible valuation of deer in the United States. Wildl Soc Bull 25:298–305
Conover M (1999) Can waterfowl be taught to avoid food handouts through conditioned food aversions? Wildl Soc Bull 27:160–166
Conover M, Pitt WC, Kessler KK, DuBow TJ, Sanborn WA (1995) Review of data on human injuries, illnesses, and economic losses caused by wildlife in the United States. Wildl Soc Bull 23:407–414
Cooper CA, Neff AJ, Poon DP, Smith GR (2008) Behavioral responses of eastern gray squirrels in suburban habitats differing in human activity levels. Northeast Nat 15:619–625
Cronin EW Jr, Sherman PW (1976) A resource-based mating system:the orange-rumped honeyguide. Living Bird 25:5–32
Crook JH, Ellis JE, Goss-Custard JD (1976) Mammalian social systems: structure and function. Anim Behav 24:261–274
Cryan PM (2008) Mating behavior as a possible cause of bat fatalities at wind turbines. J Wildl Manage 72:845–849
Cypher BL, Yahner RH, Cypher EA (1988) Seasonal food use by white-tailed deer in southeastern Pennsylvania. Environ Manage 12:237–242
Davies NB (1978) Ecological questions about territorial behaviour. In: Krebs JR, Davies NB (eds) Behavioural ecology:an evolutionary approach. Sinauer Associates, Inc, Sunderland, MA, pp 317–350
Davis DE (1967) The annual rhythm of fat deposition in woodchucks (*Maromota monax*). Physiol Zool 40:391–402
Dawkins R, Krebs JR (1978) Animal signals: information or misinformation? In: Krebs JR, Davies NB (eds) Behavioural ecology: an evolutionary approach. Sinauer Associates, Inc, Sunderland, MA, pp 282–309
Dijak WD, Thompson FR III (2000) Landscape and edge effects on the distribution of mammalian predators in Missouri. J Wildl Manage 64:209–216
Douglas RJ (1976) Spatial interactions and microhabitat selections of two locally sympatric voles, *Microtus montanus* and *Microtus pennsylvanicus*. Ecology 57:346–352
Dudley JP, Ginsber JR, Plumptre AJ, Hart JA, Campos LC (2002) Effects of war and civil strife on wildlife and wildlife habitats. Conserv Biol 16:319–329
Dumser JB (1980) The regulation of spermatogenesis in insects. Ann Rev Entomol 25:341–369

Dzialak MR, Lacki MJ, Carter KM, Huie K, Cox JJ (2006) An assessment of raptor hacking during a reintroduction. Wildl Soc Bull 34:542–547

Earle SA (1979) The Gentle whales. Nat Geogr Mag 155(1):2–17

Eckert CG, Weatherhead PJ (1987) Male characteristics, parental quality and the study of mate choice in the red-winged blackbird (*Agelaius phoeniceus*). Behav Ecol Sociobiol 20:35–42

Edmunds M (1974) Defence in animals. Longman, New York

Edwards JW, Guynn DC Jr (1995) Nest characteristics of sympatric populations of fox and gray squirrels. J Wildl Manage 59:103–110

Edwards ED, Gentili P, Horak M, Kristensen NP, Nielsen ES (1999) The cossoid/sesioid assemblage. In: Kristensen NP (ed) Lepidoptera, moths and butterflies. Walter de Gruyter, New York, pp 183–185

Eibel-Eibesfeldt I (1970) Ethology: the biology of behavior. Holt, Rinehart and Winston, New York, 530 pp

Eisenberg J (1966) The social organizations of mammals. Handbuch der Zollogie, vol 10. Walter de Gruyter & Co, Berlin, pp 1–92

Endler JA (1978) A predator's view of animal color patterns. Evol Biol 11:319–364

Etkin W (1971) Social behavior from fish to man, 3rd edn. The University of Chicago Press, Chicago, IL, 205 pp

Ewer RF (1973) The carnivores. Cornell University Press, Ithaca, NY

Field SA, Keller MA (1993) Alternative mating tactics and female mimicry as post copulatory mate guarding behavior in the parasitic wasp *Cotesia rubecula*. Anim Behav 46:1183–1189

Fisher J, Hinde RA (1949) The opening of milk bottles by birds. British Birds 42:347–357

Fisher RJ, Wiebe KL (2006) Investment in nest defense by northern flickers: effect of age and sex. Wilson J Ornithol 118:452–460

Fowler LJ, Dimmick RW (1983) Wildlife use of nest boxes in eastern Tennessee. Wildl Soc Bull 11:178–181

Fox JE (1982) Adaptation of gray squirrel behavior to autumn germination by white oak acorns. Evolution 36:800–809

Fox D (2003) More than meets the eye: behavior and conservation. Conserv Pract 4(3):20–29

Frank KD (2002) Effects of artificial night lighting on moths. In: Rich C, Longcore T (eds) Ecological consequences of artificial night lighting. Island Press, Washington, DC, pp 305–344

Franzreb KE (1983) A comparison of avian foraging behavior in unlogged and logged mixed-coniferous forest. Wilson Bull 95:60–76

Friar JL, Merrill EH, Allen JR, Boyce MS (2007) Know thy enemy: experience affects elk translocation in risky landscapes. J Wildl Manage 71:541–554

Gardner AL (1982) Virginia opossum. In: Chapman JA, Feldhammer GA (eds) Wild mammals of North America. Johns Hopkins University Press, Baltimore, MD, pp 3–36

Gates JE, Gysel LW (1978) Avian nest dispersion and fledging success in field-forest ecotones. Ecology 59:871–883

Geist V (2008) The danger of wolves. Wildl Prof 2(4):34–35

Gibbs JP, Faaborg J (1990) Estimating the viability of oven bird and Kentucky warbler populations in forest fragments. Conserv Biol 2:193–196

Gibbs JP, Wenny DG (1993) Song output as a population estimator: effect of male pairing success. J Field Ornithol 64:316–322

Giles RH Jr (1978) Wildlife management. W.H. Freeman and Co, San Francisco, CA, 416 pp

Gill FR (1990) Onithology. W.H. Freeman and Co, New York, 660 pp

Giocomo J (1998) Effects of forest openings in the contiguous forest on the reproductive success of forest songbirds. MS thesis, Pennsylvania State University, University Park

Goodenough McGuire, Wallace J, McGuire B, Wallace RA (2001) Perspectives on animal behavior, 2nd edn. Wiley, New York, 542 pp

Guidetti P (2007) Potential of marine reserves to cause community-wide changes beyond their boundaries. Conserv Biol 21:540–545

Haas CA (1995) Dispersal and use of corridors by birds in wooded patches on an agricultural landscape. Conserv Biol 9:845–854

Hahn DC, Hatfield JS (1995) Parasitism at the landscape level: cowbirds prefer forest. Conserv Biol 9:1415–1424

Harcourt AH (1991) Sperm competition and the evolution of nonfertilizing sperm in mammals. Evolution 45:314

Harris LD (1984) The fragmented forest. The University of Chicago Press, Chicago, IL

Harrison RL (2002) Estimating gray fox home-range size using half-night observation periods. Wildl Soc Bull 30:1273–1275

Harvey PH, Greenwood PJ (1978) Anti-predator defence strategies: some evolutionary problems. In: Krebs JR, Davies NB (eds) Behavioural ecology: an evolutionary approach. Sinauer Associates, Inc, Sunderland, MA, pp 129–151

Hausfater G, Hrdy SB (1984) Infanticide: comparative and evolutionary perspectives. Aldine, New York

Henner CM, Chamberlain MJ, Leopold BD, Bruger LW Jr (2004) A multi-resolution assessment of raccoon den selection. J Wildl Manage 68:179–187

Herrero S (1989) The role of learning in some fatal grizzly bear attacks on people. In: Bear-people conflicts. Proceedings of a symposium on management strategies. Northwest Territories Department of Natural Renewable Resources, Yellowknife, NT, pp 9–14

Herrero S, Smith T, DeFruyn TD, Gunther K, Matt CA (2005) From the field: brown bear habituation to people—safety, risks, and benefits. Wildl Soc Bull 33:362–373

Hill EP (1982) Beaver. In: Chapman JA, Feldhamer GA (eds) Wild mammals of North America. Johns Hopkins University Press, Baltimore

Hill GE (1993) House finch (*Carpodacus mexicanus*). In: Poole A, Gill F (eds) The birds of North America. The Academy of Natural Sciences, Philadelphia, PA

Hinde RA (1970) Animal behaviour, 2nd edn. McGraw Hill, New York

Hirth DH (1977) Social behavior of white-tailed deer in relation to habitat. Wildl Monogr 53:1–55

Hitch AT, Leberg PL (2007) Breeding distribution of North American bird species moving north as a result of climate change. Conserv Biol 21:534–539

Hobson KA, Van Wilgenburg S (2006) Composition and timing of postbreeding multispecies feeding flocks or boreal forest passerines in western Canada. Wilson J Ornithol 118:164–172

Hogg JT (1984) Mating in bighorn sheep: multiple creative male strategies. Science 225:526–529

Hoover JP, Brittingham MC (1993) Regional variation in cowbird parasitism of wood thrushes. Wilson Bull 105:228–238

Horn HS (1978) Optimal tactics of reproduction and life-history. In: Krebs JR, Davies NB (eds) Behavioural ecology: an evolutionary approach. Sinauer Associates, Inc, Sunderland, MA, pp 411–429

Husak JF, Macedonia JM, Fox SF, Sauceda RC (2006) Predation cost of conspicuous male coloration in collared lizards (*Crotaphytus collaris*): an experimental test using clay-covered model lizards. Ethology 112:572–580

Jackson WB (1982) Norway rat and allies. In: Chapman JA, Feldhamer GA (eds) Wild mammals of North America. Johns Hopkins University Press, Baltimore, pp 1077–1088

Jehl JR Jr (2006) Coloniality, mate retention, and nest-site characteristics in the semipalmated sandpiper. Wilson J Ornithol 118:478–484

Jensen RAC (1980) Cuckoo egg identification by chromosome analysis. Proc Pan Afr Ornithol Congr 4:23–25

Johnson WC, Adkisson CS (1986) Airlifting the oaks. Nat Hist 95:40–47

Johnson BE, Cushman JH (2006) Influence of a large herbivore reintroduction on plant invasions and community composition in a California grassland. Conserv Biol 21:515–526

Johnson KG, Pelton MR (1981) Selection and availability of dens for black bears in Tennessee. J Wildl Manage 45:111–119

Keister GP Jr, Anthony RG, O'Neill EJ (1987) Use of communal roosts and foraging areas by bald eagles wintering in the Klamath Basin. J Wildl Manage 51:415–420

Kilgo JC, Sargeant RA, Chapman BR, Miller KV (1998) Effect of distant width and adjacent habitat on breeding bird communities in bottomland hardwoods. J Wildl Manage 62:72–83

Kiltie RA (1989) Wildfire and the evolution of dorsal melanism in fox squirrels, *Sciurus niger*. J Mammal 70:726–739

References

Kimmel JT, Yahner RH (1991) Response of nesting goshawks to taped broadcasts. J Raptor Res 24:107–112

Kingdon J (1984) The zebra's stripes: an aid to group cohesion? In: MacDonald D (ed) The encyclopedia of mammals. Facts on File Publications, New York, pp 486–487

Kissling ML, Garton EO (2008) Forested buffer strips and breeding bird communities in southeast Alaska. J Wildl Manage 72:674–681

Kitchen DW (1984) Pronghorn. In: MacDonald DW (ed) Encyclopedia of mammals. Facts on File Publications, New York, pp 543–544

Krebs JR (1978) Optimal tactics of reproduction and life-history. In: Krebs JR, Davies NB (eds) Behavioural ecology: an evolutionary approach. Sinauer Associates, Inc, Sunderland, MA, pp 23–63

Krebs JR, Davis NB (eds) (1978) Behavioural ecology: an evolutionary approach. Sinauer Associates, Inc, Sunderland, MA

Krebs JR, Sherry DF, Healy SD, Perry VH, Vaccarino AL (1989) Hippocampal specialization of food storing birds. Proc Natl Acad Sci USA 86:1388–1392

Lack D (1947) The significance of clutch-size, I–II. Ibis 89:302–352

Lack D (1968) Ecological adaptations for breeding in birds. Methuen, London

LeBoeuf BJ (1974) Male–male competition and reproductive success in elephant seals. Amer Zool 14:163–176

Lehner PN (1982) Differential vocal response of coyotes to group howl and group yip-howl playbacks. J Mammal 63:675–679

Lehrman DS (1970) Semantic and conceptual issues in the nature–nuture problem. In: Aronson LA, Tobach E, Lehrman DS, Rosenblatt JS (eds) Development and evolution of behavior: essays in honor of T.C. Schneirla. W.H. Freeman and Co, San Francisco, CA, pp 17–52

Levine NE (1988) The dynamics of polyandry: kinship, domesticity and population on the Tibetan border. The University of Chicago Press, Chicago, IL, 309 pp

Lewis SA, Crasley CK, Rooney JA (2004) Nuptial gifts and sexual selection in *Photinus* fireflies. Integr Comp Biol 44:234–237

Lieske E, Myers RF (2004) Coral reef guide: Red Sea. HarperCollins, London

Litvaitis JA (1993) Response of early successional vertebrates to historic changes in land use. Conserv Biol 7:866–881

MacDonald DW (1976) Food caching by red foxes and some other carnivores. Zeit Fur Tierpsychol 42:170–185

Mahan CG, Yahner RH (1999) Effects of forest fragmentation on behaviour patterns in the eastern chipmunk (*Tamias striatus*). Can J Zool 77:1991–1997

Marchinton RL, Hirth DH (1984) Behavior. In: Halls LK (ed) White-tailed deer: ecology and management. Stackpole Books, Harrisburg, pp 129–168

Marti CD, Kochert MN (1995) Are red-tailed hawks and great horned owls diurnal-nocturnal dietary counterparts? Wilson Bull 107:615–628

Martin K (1998) The role of animal behavior studies in wildlife science and management. Wildl Soc Bull 26:911–920

Massey A (1988) Sexual interactions in red-spotted newt populations. Anim Behav 36:205–210

Maxson SK, Oring LW (1980) Breeding season time and energy budgets of the polyandrous spotted sandpiper. Behaviour 74:200–263

McDonough CM, Loughry WJ (2005) Impacts of land management practices on a population of nine-banded armadillos in northern Florida. Wildl Soc Bull 33:1198–1209

McFarland D (1999) Animal behaviour: psychology, ethology, and evolution, 3rd edn. Pearson/Prentice Hall, Harlow

Mech LD (1970) The wolf: the ecology and behavior of an endangered species. Natural History Press, New York

Mech LD (1984) Predators and predation. In: Halls LK (ed) White-tailed deer: ecology and management. Stackpole Books, Washington, DC, pp 189–200

Michener CD (2000) The bees of the world. Johns Hopkins University Press, Baltimore, 913 pp

Millions DG, Swanson BJ (2007) Impact of natural and artificial barriers to dispersal in the population structure of bobcats. J Wildl Manage 71:96–102

Mock DW (1984) Siblicidal aggression and resource monopolization in birds. Science 225:731–733
Moment GB (1962) Reflexive selection: a possible answer to an old question. Science 136:262–263
Morell V (1998) A new look at monogamy. Science 281:1982–1983
Morrell TE, Yahner RH, Harkness WB (1991) Factors affecting detection of great horned owls by using broadcast vocalizations. Wildl Soc Bull 19:481–488
Morse DH (1970) Ecological aspects of mixed-species foraging flocks of birds. Ecol Monogr 40:119–168
Morse DH (1980) Behavioral mechanisms in ecology. The Harvard University Press, Cambridge, MA, 383 pp
Mosillo M, Hekse EJ, Thompson JD (1999) Survival and movements of translocated raccoons in northcentral Illinois. J Wildl Manage 63:278–286
Müller-Schwarze D (1971) Pheromones in black-tailed deer (*Odocoileus hemionus columbianus*). Anim Behav 19:141–152
Nelson XJ, Jackson RR (2006) A predator from East Africa that chooses malaria vectors as preferred prey. PLoS One 1(1):e132. doi:10.1371/journal.pone.0000132
Nicolai J (1964) Der brutparasitsmus der Viduinae als ethologisches Problem: Prägungsphaenomene als faktoren der rassen-und artbildung. Z Tierpsychol 21:129–204
Nightingale B, Longcore T, Simenstad CA (2002) Artificial night lighting and fishes. In: Rich C, Longcore T (eds) Ecological consequences of artificial night lighting. Island Press, Washington, DC, pp 257–276
Norris S (2006) Evolutionary tinkering. Conserv Pract 7(3):28–34
Ofcarcik RP, Burns EE (1971) Chemical and physical properties of selected acorns. J Food Sci 36:576–578
Otsuka K, Kon M, Hidaka T (1986) The mating system of *Tokunagayusurika akamusi* (Diptera; Chironomidae): II. Experimental analysis of male mating behaviour at the resting place. J Ethol 4:147–152
Packer C, Sheel D, Pusey AE (1990) Why lions form groups: food is not enough. Am Nat 136:1–9
Parrish JK (1989) Re-examining the selfish herd: are central fish safer? Anim Behav 38:1048–1053
Partridge L (1978) Habitat selection. In: Krebs JR, Davies NB (eds) Behavioral ecology: an evolutionary approach. Sinauer Associates, Inc, Sunderland, MA, pp 351–376
Pasteur G (1982) A classificatory review of mimicry systems. Ann Rev Ecol Syst 13:169–199
Pavlov I (1927) The conditioned reflex. Oxford Press, London
Payne RB (1973) Individual laying histories and the clutch size and numbers of eggs of parasitic cuckoos. Condor 75:414–438
Payne RB (1977) The ecology of brood parasitism in birds. Ann Rev Ecol Syst 8:1–28
Penny D, Ortuño VM (2006) Oldest true orb-weaving spider (Araneae: Araneidae). Biol Lett 2:447–450
Perry HR Jr (1982) Muskrats. In: Chapman JA, Feldhammer GA (eds) Wild mammals of North America. Johns Hopkins University Press, Baltimore, MD, pp 282–325
Persons MH, Uetz GW (2005) Sexual cannibalism and mate choice decision in wolf spiders: influence of male size and secondary sexual characters. Anim Behav 69:83–94
Peters RP, Mech LD (1975) Scent-marking in wolves. Am Sci 63:628–637
Peterson RO, Page RE (1988) The rise and fall of the Isle Royale wolves, 1975–1986. J Mammal 69:89–99
Polak M, Wolf LL, Starmer WT, Barker JSF (2001) Function of the mating plug in *Drosophila hibisci* Bock. Behav Ecol Sociobiol 9:196–205
Pough FH (1976) Multiple cryptic effects of cross-banded and ringed patterns of snakes. Copeia 1976:834–836
Pough FH, Janis CM, Heister JB (2002) Vertebrate life, 6th edn. Prentice Hall, Upper Saddle, NJ
Probst JR, Weinrich J (1993) Relating Kirtland's warbler population to changing landscape composition and structure. Landsc Ecol 8:257–371
Reavis RH (1997) The natural history of a monogamous coral-reef fish, *Valenciennea strigata* (Gobiidae). 2. Behavior, mated fidelity, and reproductive success. Environ Biol Fishes 49: 247–257

References

Rich C, Longcore T (eds) (2006) Ecological consequence of artificial night lighting. Island Press, Washington, DC

Risenhoover KL, Peterson RO (1986) Mineral licks as a sodium source for Isle Royale moose. Oecologia 71:121–126

Rodewald AD, Yahner RH (1999) Effects of forest management and landscape composition on woodland salamanders. Northeast Wildl 54:45–54

Rodewald AD, Yahner RH (2000) Influence of landscape and habitat characteristics on ovenbird pairing success. Wilson Bull 112:238–242

Rogers LL (1980) Inheritance of coat color and changes in pelage coloration in black bears in northeastern Minnesota. J Mammal 61:324–327

Rogers LL (1986) Effects of translocation distance on frequency of return by adult black bears. Wildl Soc Bull 14:76–80

Rohnke AD, Yahner RH (2008) Long-term effects of wastewater irrigation on habitat and a bird community in central Pennsylvania. Wilson J Ornithol 120:146–152

Rollfinke BF, Yahner RH (1990) Effects of time of day and season on winter bird counts. Condor 92:215–219

Rollfinke BF, Yahner RH (1991) Flock structure of wintering birds in an irrigated mixed-oak forest. Wilson Bull 103:282–285

Rusch DH, Meslow EC, Doerr PD, Keith LB (1972) Response of great horned owl populations to changing prey populations. J Wildl Manage 36:282–296

Sajer BF (1953) Science and human behavior. Free Press/Macmillan, New York

Sargent TD (1969) Behavioural adaptations of cryptic moths. III. Resting attitudes of two bark-like species, *Melanolophia canadaria* and *Cotocala ultronia*. Anim Behav 17:670–672

Sauer JR, Hines JE, Fallon J (2008) The North American breeding bird survey, results and analysis 1966–2007. Version 5.15.2008. USGS Patuxent Wildlife Research Center, Laurel, MD

Schaller GB (1967) The deer and the tigers study life in India. The University of Chicago Press, Chicago, IL

Scott DP, Yahner RH (1989) Winter habitat and browse use by snowshoe hares, *Lepus americanus*, in a marginal habitat in Pennsylvania. Can Field Nat 103:560–563

Serfass TL, Brooks RP, Rymon LM (1993) Evidence of long-term survival and reproduction by translocated river otters, *Lutra canadensis*. Can Field Nat 107:59–63

Servello FA, Hellgreen EC, McWilliams SR (2005) Techniques for wildlife nutritional ecology. In: Braun CE (ed) Techniques for wildlife investigations and management. Port City Press, Baltimore, pp 554–590

Shaffer LL (1995) Pennsylvania amphibians and reptiles. Pennsylvania Fish and Boat Commission, Harrisburg

Sherman P (1977) Nepotism and the evolution of alarm calls. Science 197:1246–1253

Shew JJ (2006) American crow caches rabbit kits. Wilson J Ornithol 118(7):572–573

Shine R (1988) Parental care in reptiles. In: Gans C, Huey RB (eds) Biology of the reptilia, vol 16. Alan R. Liss, Inc, New York, pp 275–330

Smith NG (1966) Evolution of some arctic gulls (Larus): an experimental study of some isolating mechanisms. Ornithol Monogr 4:1–99

Smith CC (1968) The adaptive nature of social organization in the genus of tree squirrels Tamiasciurus. Ecol Monogr 38:31–63

Smith TA, Herrero S, Debruyn TD, Wilder JM (2008) Efficacy of bear deterrent spry in Alaska. J Wildl Manage 72:640–645

Steele MA, Wauters LA, Larsen KW (2005) Selection, predation and dispersal of seeds by tree squirrels in temperate and boreal forests: are tree squirrels keystone granivores? In: Forget J-M, Forget P-M, Lambert JE, Hulme PE, Vander Wall SB (eds) Seed fate: predation, dispersal, and seedling establishment. CAB International Wallingford, Oxon, pp 205–221

Stewart A (2004) The earth moved: on the remarkable achievements of earthworms. Algonquin Books, Chapel Hill, NC

Stober JM, Krementz DG (2006) Variation in Bachman's sparrow home-range size at the Savannah River Site, South Carolina. Wilson J Ornithol 118:138–144

Stout GG (1982) Effects of coyote reduction on white-tailed deer productivity on Fort Sill, Oklahoma. Wildl Soc Bull 10:329–332

Sutcliffe OL, Thomas CD (1996) Open corridors appear to facilitate dispersal by ringlet butterflies (*Aphantopus hyperantus*) between woodland clearings. Conserv Biol 10:1359–1365

Sweanor LL, Logan KA, Bauer JW, Millsap B, Boyce WM (2008) Puma and human spatial and temporal use of a popular California State park. J Wildl Manage 72:1076–1084

Swihart RK, Yahner RH (1982) Eastern cottontail use of fragmented farmland habitat. Acta Theriol 27:257–273

Thomas KR (1974) Burrow systems of the eastern chipmunk (*Tamias striatus* pipilans Lowery) in Louisiana. J Mammal 55:454–459

Thompson WL, Yahner RH, Storm GL (1990) Winter use and habitat characteristics of vulture communal roosts. J Wildl Manage 54:77–83

Thorndike EI (1898) Animal intelligence: an experimental study of the associative process in animals. Psychol Monogr 2:1–109

Thurber JM, Peterson RO (1991) Changes in body size associated with range expansion in the coyote (*Canis latrans*). J Mammal 72:750–755

Tinbergen N (1963) On aims and methods of ethology. Z Tierpsychol 20:410–433

Todd AW, Keith LB, Fischer CA (1981) Population ecology of coyotes during a fluctuation of snowshoe hares. J Wildl Manage 45:629–640

Toner JB, Adler NT (1985) Potency of rat ejaculations varies with their order and with male age. Physiol Behav 35:113–115

Trivers RL (1974) Parent-offspring conflict. Am Zool 14:249–264

Tsubaki Y, Siva-Jothy MT, Tomohiro Ono O (1994) Re-copulation and post-copulatory mate guarding increase immediate female reproductive output in the dragonfly *Nannophya pygmaea* Rambur. Behav Ecol Sociobiol 35:219–225

Vander Wall SB (1990) Food hoarding in animals. The University of Chicago Press, Chicago, IL

Vaughan TA, Ryan JM, Czaplewski NJ (2000) Mammalogy. Saunders, New York, 565 pp

Vecellio GM, Yahner RH, Storm GL (1994) Crop damage by deer at Gettysburg Park. Wildl Soc Bull 22:89–93

Verner J (1977) On the adaptive significance of territoriality. Am Nat 111:769–775

Vickery PD (2002) Effects of the size of prescribed fire on insect predation of north blazing star, a rare grassland perennial. Conserv Biol 16:413–421

Vilà C, SavolainAen P, Maldonado JE, Amorim IR, Rice JE, Honeycutt RL, Crandall KA, Lundeberg J, Wayne RK (1997) Multiple and ancient origins of the domestic dog. Science 276:1687–1689

Wahaj SA, Place NJ, Weldele ML, Glickman SE, Holekamp KE (2007) Siblicide in the spotted hyena: analysis with ultrasonic examination of wild and captive individuals. Behav Ecol 18:974–984

Walther FR (1969) Flight behaviour and avoidance of predators in Thomson's gazelle (*Gazella thomsonsi* Guenther 1884). Behaviour 34:184–221

Weaver HW, Anderson JT, Edwards JW, Dotson TL (2004) Black bear response to physical and auditory conditioning techniques in southern West Virginia. Northeast Wildl 58:23–33

Weidensaul S (2007) The last gladiators. Conserv Mag 8(3):19–21

Weldon A (2006) How corridors reduce indigo bunting nest success. Conserv Biol 20:1300–1305

Welty JC (1982) The life of birds, 3rd edn. Saunders, Philadelphia, PA

Whitaker JO Jr (1972) Food habits of bats from Indiana. Can J Zool 50:877–883

White TH Jr, Bowman JL, Jacobson HA, Leopold BD, Smith WP (2001) Forest management and female black bear denning. J Wildl Manage 65:34–40

Whitman KL, Starfield AM, Wuadling H, Packer C (2007) Modeling the effects of trophy selection and environmental disturbance on simulated population of African lions. Conserv Biol 21:591–601

Wilcox BA, Murphy DD (1985) Conservation strategy: the effect of fragmentation on extinction. Am Nat 125:879–887

Wilson EO (1971) The insect societies. The Harvard University Press, Cambridge, MA

Wilson EO (1975) Sociobiology: the new synthesis. The Harvard University Press, Cambridge, MA, 697 pp

References

Wise SE, Buchanan BW (2002) Influence of artificial illumination on the nocturnal behavior and physiology of salamanders. In: Rich C, Longcore T (eds) Ecological consequences of artificial night lighting. Island Press, Washington, DC, pp 221–251

Withers GS, Fahrbach SE, Robinson GE (1993) Selective neuroanatomical plasticity and division of labor in the honeybee. Nature 364:238–240

Woolfenden GE (1973) Nesting and survival in a population of Florida scrub jay. Living Bird 12:25–49

Woolfenden GE, Fitxpatick JW (1984) The Florida scrub jay. Princeton University Press, Princeton, NJ

Wright AL, Yahner RH, Storm GL (1986) Habitat use and abundance of wintering vultures at Gettysburg, Pennsylvania. J Raptor Res 20:102–107

Wynne-Edwards WC (1962) Animal dispersion in relation to social behavior. Hafner, New York

Yahner RH (1975) The adaptive nature of scatter hoarding in the eastern chipmunk. Ohio J Sci 75:176–177

Yahner RH (1978a) Burrow system and home range use by eastern chipmunks (*Tamias striatus*): ecological and behavioral considerations. J Mammal 59:324–329

Yahner RH (1978b) The adaptive nature of the social system and behavior in the eastern chipmunk, *Tamias striatus*. Behav Ecol Sociobiol 3:397–427

Yahner RH (1978c) The sequential organization of behavior in *Tamias striatus*. Behav Biol 24:229–243

Yahner RH (1978d) Seasonal rate of vocalizations in eastern chipmunks. Ohio J Sci 78:301–303

Yahner RH (1979) Temporal patterns in the male mating behavior of captive Reeve's muntjacs (*Muntiacus reevesi*). J Mammal 60:560–567

Yahner RH (1980a) Time budgets in captive Reeve's muntjacs (*Muntiacus reevesi*). Appl Anim Ethol 6:277–284

Yahner RH (1980b) Barking in a primitive ungulate, *Muntiacus reevesi*: function and adaptiveness. Am Nat 114:157–177

Yahner RH (1980c) Burrow system use by red squirrels. Am Midland Nat 103:409–411

Yahner RH (1983) Seasonal dynamics, habitat relationships, and management of avifauna associated with farmstead shelterbelts. J Wildl Manage 47:85–104

Yahner RH (1984) Avian use of nest boxes in farmstead shelterbelts. Minnesota Acad Sci 49:18–20

Yahner RH (1987) Use of even-aged stands by winter and spring bird communities. Wilson Bull 99:218–232

Yahner RH (1988) Changes in wildlife communities near edges. Conserv Biol 2:333–339

Yahner RH (1997a) Long-term dynamics of bird communities in a managed forested landscape. Wilson Bull 109:595–613

Yahner RH (1997b) Historic, present, and future status of Pennsylvania vertebrates: some issues of conservation concern. J PA Acad Sci 71:47–51

Yahner RH (1999) Edge use by butterfly communities in agricultural landscapes. Northeast Wildl 54:13–24

Yahner RH (2000) Eastern deciduous forest: ecology and wildlife conservation, 2nd edn. University of Minnesota Press, Minneapolis

Yahner RH (2001) Fascinating mammals: conservation and ecology in the mid-eastern States. University of Pittsburgh Press, Pittsburgh, PA

Yahner RH (2004a) The changing forest of Pennsylvania: potential implications to wildlife. Northeast Wildl 58:35–47

Yahner RH (2004b) The taxonomy-conservation gap: rejuvenating conservation careers and mentoring students. Conserv Biol 18:6–7

Yahner RH, Mahan CG (1997) Behavioral considerations in fragmented landscapes. Conserv Biol 11:569–570

Yahner RH, Mahan CG (2002) Animal behavior in fragmented landscapes. In: Gutzwiller KJ (ed) Applying landscape ecology in biological conservation. Springer, New York, pp 266–285

Yahner RH, Morrell TE (1991) Depredation of artificial avian nests in irrigated forests. Wilson Bull 103:113–117

Yahner RH, Ross BD (1995) Distribution and success of wood thrush nests in a managed forested landscape. Northeast Wildl 52:1–9

Yahner RH, Scott DP (1988) Effects of forest fragmentation on depredation of artificial nests. J Wildl Manage 52:158–161

Yahner RH, Svendsen GE (1978) Effects of climate on the circannual rhythm of the eastern chipmunk, *Tamias striatus*. J Mammal 59:109–117

Yahner RH, Wright AL (1985) Depredation on artificial avian ground nests: effects of edge and age of small aspen plots. J Wildl Manage 49:510–515

Yahner RH, Hutnik RJ, Liscinsky SA (2003) Long-term trends in bird populations on an electric transmission right-of-way. J Arboric 29:156–163

Young GI, Yahner RH (2003) Distribution of and microhabitat use by woodland salamanders along forest-farmland edges. Can Field Nat 117:19–24

Zug GR, Vitt LJ, Calwell JP (2001) Herpetology, 2nd edn. Academic Press, New York

Index

A
Acuity, visual, 102
African lion *(Panthera leo)*
   dominance status, 99
   female, 27
   infanticide, 22
   prey size, 10
   pride size, 19
   and spotted hyenas, 88
   subordinate male, 97
   Tanzania, 19
Aggression
   black bear, 53
   dominance, measurement, 98
   flickers, 84
   humans, 100
   inappropriate, 100
   individuals decline, 92
   interactions, 98–99
   northern mockingbird, 60
   overt, 84
   prospective mates, 25
   squirrels, 90
   territorial defense, 84, 89
Alarm
   call
      axis deer, 123
      birds, 64, 124
   signal
      forest deer, 121–122
      olfaction, 115
   substance, 63
Altricial and precocial young
   bird and mammals, 31
   development, 31
   dichotomy, 31
   filial imprinting, 31–32
   imprinting-like processes, mammals, 32
   kangaroos, 32–33
   parental care, 31
   propagation programs, problem, 32
   sexual imprinting, 32
Altruism and parental care
   alarm calling, 27
   Florida scrub jays, 27
   helpers/altruistic animals, 27
Animal
   aerial, daily torpor, 142–143
   behaviors, "Skinner box,"
   learning, social unit, 10
   recognize neighbors, 87
   terrestrial, 129
   threat displays and decision making, 105
   weaponry, 62–64
Antler evolution, 107–108
Auditory communication
   barking
      domestication, domestic dogs, 122–123
      forest deer, 121–122
   birds
      identification, songs, 126
      mimicry, 127
      song repertoires, 126
      vocalizations, 124–126
   sound effects, terrestrial animals, 129
   sound-producing mechanisms
      fishes, 128
      mammals and birds, 127–128
      tourist and research vehicles, 128
   vocalizations, wildlife
      dominance signals, 124
      gray wolves and coyotes, 123
      male eastern chipmunks, 124
      younger animal, 123–124

B

Bark
  domestic dogs
    gray wolves and human groups, 122–123
    mitochondrial DNA sequencing, 122
  forest deer, 121–122
Bates, H.W., 59
Batesian mimicry, 59
Behavioral ecology, 6
  Behaviors. *See also* Genetics and mechanisms, behavior
    displacement, 106
    diversity, 5
    play, 135
    "risk-free", 104
    wildlife
      American psychologists, 4
      Carnis familiaris dog salivates, 3
      European ethologists, 4
      FAP, 4
      integrates, 4
      management, 1–2
      nuisance, 3
      "Skinner box", 2–3
Birds
  brown-headed cowbird, 29–31
  echolocation, 131–132
  identification, songs, 126
  mimicry, 127
  song repertoires, 126
  vocalizations
    alarm and contact calls, 124
    duets, 125–126
    ovenbird, 124–125
    playback studies, 125
Body fat
  black bear, 139, 141
  ground squirrel, 140
  woodchuck, 140–141
Brood parasitism and parental care
  adaptations, 28–29
  cowbirds, 30
  intraspecific parasitism, 28
  occurs, 28
  strategies, 28
  taxonomic groups, 31
  waterfowl, nest, 28
Browse, 153, 154

C

Cavity
  primary nester, 74
  tree, 73
  use, 74
Central place foraging (CPF)
  American beaver, 47
  eastern chipmunk, 47
  larder and scatter, 46–47
  migratory wildebeest, 48
  patchs, 47–48
  right time, 48
  strategies, 46
  types, 46
Chemical
  biochemical factor affects, 13
  castor, 119
  noxious, 63
  olfactory, 114
  Schreckstoff, 115
  semiochemicals, 114
  undecane, 115
  warfare, 62
  white-tailed deer, role, 116
Commensal mice, 65
Communication
  animals, 101
  antlers and horns, evolution
    Bovidae family, 108
    hypotheses, 107–108
    sex-specific reasons, 108
    ungulates, 107
  artificial night lighting, 108
  auditory, 101
  echo, 131
  echolocation
    bats and birds, 131–132
    shrews, 132
  ecological light pollution, wildlife
    birds, 109–110
    fishes, 110–111
    insects, 111
    nocturnal mammals, 109
    sea turtles, 110
    snakes and toads, 110
  electrical, 136–137
  evolutionary origin, displays, 106
  Homo habilis, 101, 102
  olfactory (*see* Olfactory communication)
  pinnipeds, toothed whales and African elephants, 132–133
  play behavior, 135
  seismic, 135–136
  tactile, 134
  tusks, walrus, 106–107
  visual
    birds, 102
    roles, 103–104

Index

signal, 102
tapetum lucidum, 102
threat displays, 104–105
Comparative psychology.
    See Psychology *vs.* ethology
Competition
  interspecific
    ants and pigs, 153–154
    black bears, 154
    food, 152–153
    gray and fox squirrels, 154
    sika deer and cane toads, 153
    vandalism, 152
    vultures, 152
  macaques, 151–152
  mates, 151
  types, 151
Conditioning
  Carnis familiaris dog salivates, 3
  FAP, 4
  integrates, 4
  nuisance wildlife, 3
Core area
  chipmunk, 84
  definition, 80
Corridors
  animals use, 37
  dimensions and composition, 37
  foraging strategy, species, 38
  greenside darter, 39
  habitat patches, isolation, 37
  populations, 39
  quality, distant habitat, 38
  scientists to "sell", 37
  sell to public, 37
  vegetation, 38
  width and character, 38–39
  wildlife, birds, 38
  woodlots isolation, 37–38
Countershading, 55–56
Courtship
  function, 25
  males, 20
  and mating systems, 15
Crypsis
  coloration, 57
  predator, 56, 58
Cues
  navigation
    homing pigeons, 149
    olfaction, 150
    visual, 149
  use, 99
  visual/auditory, 19, 59, 99, 109

D

Daily torpor, aerial animals
  black-capped chickadees, 142
  ruby-throated hummingbird, 142–143
  vespertilionid bats, 143
Darwin, C., 6
Den
  communal, 91
  sites, 73, 74
  temperature, 91
  tree cavities, 154
Dispersal
  being philopatric, 36
  definition, 35
  gypsy moth, forest pest, 36, 37
  timing, 35
Domestication, domestic dogs
  gray wolves and human groups, 122–123
  mitochondrial DNA sequencing, 122
Dominance
  hierarchies (*see* Dominance hierarchies)
  signals, 124
  territory, 83
Dominance hierarchies
  advantages, 97
  aggression, 100
  domestic chickens, 95, 96
  eastern chipmunks, 96
  maintenance and establishment
    aggressive interactions, 98–99
    body size, 99
    eastern chipmunks and
      copperheads, 99–100
    raptor species, 100
  measurement, 98
  pecking order, 95
  resource and survival, 100
  subordinate animals and grouping, 97–98
  territoriality and individual distance, 95
Dormancy
  vs. migration, 145–146
  winter strategies, 139
Duet
  "duet code", 125
  marsh wrens, 126
  use, 125

E

Echo, 131, 132
Echolocation
  bats
    little brown myotis, 131–132
    primary and secondary function, 131

Echolocation (*cont.*)
    tragus, 132
    birds, 132
    shrews, 132
Eisenberg, J., 15
Electrical communication
    Eimer's organs, 137
    fishes, 136
    honeybee scouts, 134
    mammals, 136–137
Electricity, 136
Electroreception, 136–137
Energy stores, migration, 147
Exploitative competition, 151, 154

F
FAP. *See* Fixed action pattern
Fight, 5
Fixed action pattern (FAP), 4
Flehmen, 114
Floater
    population, 90
    postbreeding flocks, 8
    presence, 91
    territory, 91
Food-acquisition systems
    bees, 42
    conservation and warfare, 53
    CPF and hoarding, food, 46–48
    feeding, 41
    food capture and process, organisms, 41
    foraging and group life, 49–50
    herbivores, 42
    leaf cutter ants, 42
    optimal foraging theory (*see* Forage)
    predation and competition, 48–49
    prey
        distribution and predation, 50–51
        humans, 51–53
        rat race and wildlife, 42–43
    white-tailed deer, 42
Forage
    African hunting dog, 49
    arthropods, 50
    CPF (*see* Central place foraging (CPF))
    food habit and predator–prey studies, 49–50
    foraging strategies, 50
    gypsy moth defoliation, 50
    marine reserves, 49
    optimal foraging theory
        common raccoons, 45
        functional response, 44–45
        moose, 45
        numerical response, 45
        predation and competition, 48–49
        predators, 43–44
        predictions stem, 44
        South Carolinian and Georgian wetlands, 45
        wood storks, 45, 46

G
Generalist *vs.* specialist
    adaptations, reptiles, 74
    black bear, 73
    dichotomy, 73
    downy woodpecker, 74
    nest boxes, 74
    northern flying squirrels, 74
    resources, 74–75
    songbird boxes, 74
    tree dens, use, 73
Genetics and mechanisms, behavior
    cultural transmission, 10–11
    diversity, 5
    hormones and proximate factors, 12–13
    nervous system and biochemistry, 13
    social organization/system, 6
    units (*see* Social units)
    sociobiology and behavioral ecology, 6
    ultimate *vs.* proximate factors, wildlife, 11–12

H
Habitat selection
    factors affects
        acid deposition, 70
        ambient temperature, 69–70
        humans, 67–68
        long-tailed weasel, 68–69
        population density, 70–71
        song rates estimate, birds, 68
        taxa, 68
        tradition, 71–72
        woodland salamanders, 68
    generalist *vs.* specialist, 73–75
    learning, young animals, 72
    quantify problems, 75
    and urban wildlife, 66–67
    vs. testable-habitat use, 66
    vs. uses, 65
    and wildlife recolonization, 67

Index

Habitat use
  Appalachian Mountains, 65
  commensal mice, 65
  correlations, 66
  flying squirrels, 65
  habitat selection, species, 66
  Norway rat, 65
Heartbeat, 101, 141
Helper
  advantages, 27
  breeding pairs, 27
  brown-headed nuthatch, 27
Hertz (Hz), 131–133, 136
Hibernation
  bats, 141
  desert tortoise, 140
  ground squirrels, 140
  woodchuck, 140–141
Hierarchy, dominance. *See* Dominance hierarchies
Hippocampus, 13
Hoarding and CPF
  American beaver, 47
  eastern chipmunk, 47
  larder and scatter, 46–47
  migratory wildebeest, 48
  patchs, 47–48
  right time, 48
  strategies, 46
  types, 46
Home range
  definition, 77
  dispersion pattern, relationship, 77
  quantification, 79
  shape
    core area, 80
    coyotes, 80–81
    eastern cottontail, 80, 81
    WNV, 81–82
  size, 79–80
Homing
  pigeons, 114, 149
  wildlife
    common raccoons, 78–79
    definition, 78
    nuisance black bear, 78
    red-legged frogs, 78
Homo habilis, 101, 102
Homo neanderthalensis, 101
Horn
  bighorn sheep, 147
  evolution, 107–108
  large and near, 62
  owl, 44
  species, 63
Howl
  group, 85
  intrapack, 123
  yip-howl, 85, 123
Human
  communication, 101
  perspiration and blushing, 106
  prey (*see* Prey)

I

Individual distance
  adult males and females, 92
  brown-headed nuthatches, 92
  chipmunks, 92
  defense, 91
  mobile territories, 93
Infanticide
  male lions, 99
  olfaction, 117
  role, 1
  species, 22
Infidelity, 17
Infrasound, 133
Interference competition, 151, 152
Interspecific competition
  ants and pigs, 153–154
  black bears, 154
  food, 152–153
  gray and fox squirrels, 154
  sika deer and cane toads, 153
  vandalism, 152
  vultures, 152
Intra-specific
  home range, 79
  parasitism, 28

J

Jacobson's organ. *See* Vomeronasal organ

L

Lack, D., 26
Lactation
  mammals, 26–27
  metatherians, 32
Landscape linkages
  megacharismatic species, 39–40
  spectacular migrations, 40
  swamp, 39
Larder hoarding. *See* Hoarding
Lateral lines, 136

Learning
  cultural transmission, 10–11
  navigation, 150
Lek
  behavior, 18, 19
  benefits, 19

M
Magnetism, 149
Management
  habitat, 50
  timber practices, 154
  wildlife and behavior (*see* Wildlife)
Marsupial, 32, 43
Mate-acquisition and parental-care systems
  courtship and mating systems, 15
  mate choice, 22–24
  mating and mechanisms, interference, 22
  monogamy, 15–17
  polygamy, 17–20
  promiscuity and mating systems, 20–21
Mating chase, 96, 98–99
Mating systems
  and courtship, 15
  monogamy (*see* Monogamy)
  and parental care
    altricial and precocial young, 31–32
    and altruism, 27
    and brood parasitism, 28–29
    brown-headed cowbird, 29–31
    courtship, 25
    dispersal process, 33
    lactation, mammals, 32
    locomotion, 32–33
    parents, 25–27
    pouches, 32–33
    precocial young *vs.* altricial, 32–33
  promiscuity (*see* Promiscuity)
Matriarchal social units, 9
  Megacorridors. *See also* Landscape linkages
    definition, 39
    mammals, 40
Melanism, 57
Microphone, 131
Migration
  advantages, 145
  birds, 145
  disadvantages, 145–146
  distances
    altitudinal migration, 146–147
    mountain sheep, 146, 147
    Serengeti wildebeest, 146
  energy storage, 147
  timing
    arrival dates, 148
    black brant, 148
    cliff swallows, 147
    hawks and warblers, 148–149
    restlessness, 147–148
Mimicry
  alligator snapping turtles, 60
  Batesian mimicry, 59
  behavior, 60
  brood parasites, 60–61
  butterflies, 61
  forms, 59
  harmless insects, 60
  lizards and snakes, 61
  mockingbird, 60
  Müllerian mimicry, 59
  sea cucumbers, 61
  spotted hyenas, behavior, 60
Monogamy
  infidelity, 17
  Kirk's dik-dik, 16
  male, 16
  perennial and seasonal, 16–17
  sexes, species, 15–16
  species, 21
  yellow warbler, 16
Müllerian mimicry, 59
Mushroom bodies, 13

N
Navigation
  cues, 149–150
  learning, 150
  routes, 149

O
Observable response, 101
Olfaction
  alarm signals, 115
  infanticide, 117
  wildlife species, 113–114
Olfactory communication
  alarm signals, 115
  beaver and mud piles
    castor and anal glands, 118–119
    colony, 118
  buck
    rubs, 115
    scrapes, 116
  cheek rubbing, 116–117

Index

infanticide, 117
mammals, 116
raccoons, 116, 117
scent-marking, canids, 117–118
scent stations, mammalian research, 119–120
smell *vs.* taste, 113
wildlife species
  allomones and pheromones, 114
  birds, 113–114
  fish lures, 113, 114
  sharks and salmon, 113
  vomeronasal/Jacobson's organ, 114
Optimal foraging theory
  common raccoons, 45
  functional response, 44–45
  moose, 45
  numerical response, 45
  predation and competition, 48–49
  predators, 43–44
  predictions stem, 44
  in South Carolinian and Georgian wetlands, 45
  wood storks, 45, 46
Orientation, 149, 150

P

Parental care. *See* Mate-acquisition and parental-care systems
Pavlov, I, 3
Pecking order, 95
Philopatry, 35–36
Polyandry
  fraternal, humans, 20
  spotted sandpiper, 20
Polygamy
  American jaçana, 20
  benefits of leks, 19
  extra-pair matings, 19
  fur seals, 17, 18
  harvest effects, 19
  humans, 20
  hunters, 19
  lek behavior, 18
  mammals, 18
  species, 17
  spotted sandpiper, 20
  types, 17
Polygyny
  bird species, 17
  females lekking behavior, 19
  harvest effects, 19
  humans, 20
  lek behavior, 18

Polymorphism, types, 57
Precocial young. *See* Altricial and precocial young
Predation
  and competition, 48–49
  efficiency, 44
  less-direct adaptations
    animals, 56
    black bears, 57–58
    cephalopods, 56
    crypsis, 58
    fox squirrels, 56, 57
    jellyfish, 56
    jumping spiders, 57
    marsh birds, 55
    monarchs, adults, 55
    peppered moth, 57
    predation, 58
    prey species, 55–56
    water snakes, 57
  mimicry (*see* Mimicry)
  playing possum and enhancement, 61–62
  and prey distribution, 50–51
  warning coloration (*see* Warning)
  weaponry animals, 62–64
Predator
  attack large prey, 43
  determine, 48
  and food habit, 49–50
  hunting cryptic prey, 43
  prey, 43–44
  reef fish, 48
  sit-and-wait, 42–43
Prey
  American crow, 45
  distribution and predation
    browsing, deer, 50–51
    elk, 51
    grazing, deer, 51
    humans
      bear attacks, 52–53
      and mountain lion, 51–52
      in North America, 51
      red-pepper sprays, 53
  locates, 45
  rat race and wildlife
    chasing prey, 43
    humpback whale, 42
    hunting cryptic prey, 43
    orb-weaving spiders, 42
    sensors, predators, 43
    sit-and-wait predators, 42–43
  size, 44
  use, 48–49

Pride, 10, 19, 97
Promiscuity and mating systems
　eastern chipmunks, 20–21
　giant water bug, 21
　harem-forming species, 21
　hermaphroditism, 21
　internally fertilized species, 21
　monkey species, 21
　monogamous species, 21
　protogynous hermaphroditism, 21
Psychology *vs.* ethology
　behavior and wildlife management, 1–2
　wildlife behaviors
　　American psychologists, 4
　　Carnis familiaris dog salivates, 3
　　European ethologists, 4
　　FAP, 4
　　integrates, 4
　　nuisance, 3
　　"Skinner box", 2–3

R
Recolonize, wildlife, 67
Resource
　Commensal, 65
　environment, 6
　food, 68
　limiting, 21
　parents, 25–26
　polygyny, 19
　rich-food, 47
　role (*see* Generalist *vs.* specialist)
　Wisconsin Department of Natural Resources, 67
Ripples, 136
Rogers, L.L., 78
Royal catchfly, 47

S
Satter hoarding. *See* Hoarding
Scent-marking
　canids, 117–118
　gray wolves, 116
　single, 85
Scent station, mammalian research
　description, 119
　investigators, 120
　nest predation, birds, 119
Seasonal torpor. *See* Winter strategies
Seismic communication
　lateral lines, 136
　snakes, 136
　white-lipped tree frog, 135
Sensory perception, 101
Skinner, B.F., 2–3
"Skinner box", 2–3
Social system, 6
Social units
　animals
　　distribution, habitats, 6, 7
　　learn, 10
　bald eagles, 8
　bird species, northern boreal forests, 8
　brown bears, western states, 7
　determination, 8
　matriarchal, 9
　plasticity/flexibility, size, 9–10
　semipalmated sandpiper, 8
　spatial distribution/dispersion pattern, types, 7
　turkey and black vultures, 8
　vs. aggregation, 7
　winter and nonwinter units, 8–9
Sociobiology, 6
Spacing mechanisms
　defense, individual distance
　　brown-headed nuthatches, 92
　　communal denning, 91
　　description, 91
　　eastern chipmunks, 92
　　European rabbit and chaffinch, 92
　　mobile territories, 93
　　winter, 91
　neighbor recognition, 87
　territoriality
　　behavior, 83
　　evolution, 88–89
　　interspecific, 90–91
　　marine iguana, 87
　　North American red squirrels, 87
　　ontogeny, 87
　　territory, 83–84
　territory
　　defense, 84–86
　　definition, problems, 86–87
　　size, 88
Specialist. *See* Generalist *vs.* specialist
Stun, 133

T
Tactile communication, 134, 135
Tail flagging, 103–104
Territoriality
　behavior, 83
　evolution
　　food, role, 89

songbirds and red squirrels, 89
wildlife, 88
interspecific
    description, 90
    floater population, 90–91
    predation, 90
marine iguana, 87
North American red squirrels, 87
ontogeny, 87
territory, 83–84
Territory
cats, 83–84
defense
    American beaver, 86
    deer, 86
    northern flicker, 84, 85
    overt aggression, 84
    scent-marks, 85
    trogons, 84–85
    wolf, 85–86
definition, 83, 86–87
size, 88
vs. home range, 83
Thorndike, E.I., 2
Threat
actual chases, 84
displays and decision types, 105
and fights, 17
northern spotted owl, 74
signals, 99
status, 78
visual dsplays, 104–105
Threat displays, animals
types and decision making, 105
visual, 104–105
Tinbergen, N., 4
Trivers, R.L., 26
Tusk, 106–107

U
Ultrasound
    echo, 131
    echolocation
        bats and birds, 131–132
        shrews, 132
Urbanization, 39, 82

V
Vandalism, 152
Vision
    birds, 102
    rod-dominated, 109
Visual communication
    birds, 102
    roles
        baboons and cavity nesting species, 104
        parent-young bonds, 104
        tail flagging, 103–104
        white-tailed deer, 103
    signal, 102
    tapetum lucidum, 102
    threat displays
        spatial and temporal contexts, 104–105
        types and decision making, 105
Vomeronasal organ, 114

W
Warning coloration
    aposematism, 58–59
    stinkpot family, 59
    striped skunk, 59
    tropical frogs, 58
Water drainage, 80
West Nile virus (WNV), 81–82
Wildlife. *See also* Auditory communication
    and behavior management, 1–2
    ecological light pollution, 109–111
    homing, 78–79
    lek behavior, 18
    olfaction, 113–114
    prey rat race, 42–43
    ultimate *vs.* proximate factors, 11–12
    urban, 66–67
Winter lethargy
    black bear, 139, 141
    den sites, 73
    tree cavities, 73
Winter strategies
    bats, 139
    dormancy, 139
    hibernation
        bats, 141
        desert tortoise, 140
        ground squirrels, 140
        woodchuck, 140–141
    lethargy, black bear, 139, 141
    torpor
        aerial animals, 142–143
        Eastern chipmunks, 142
WNV. *See* West Nile virus
Wood ibis, 46